D1250725

Geometry and Induction

GEOMETRY AND INDUCTION

Jean Nicod

Containing
GEOMETRY IN THE SENSIBLE WORLD
and
THE LOGICAL PROBLEM OF INDUCTION
with prefaces by
Roy Harrod, Bertrand Russell
and André Lalande

UNIVERSITY OF CALIFORNIA PRESS
Berkeley and Los Angeles
1970

UNIVERSITY OF CALIFORNIA PRESS
Berkeley and Los Angeles, California

First English translation 1930
New translation by
John Bell & Michael Woods, 1970

Standard Book Number 520-01689-0
Library of Congress Catalog Card Number: 70-107149

Printed in Great Britain

Preface

Some years ago I suggested to Lord Russell that a new English edition of Jean Nicod's two classic essays was desirable. He cordially agreed, and offered a generous donation towards the cost. This was supplemented by another donor, who wishes to remain anonymous.

The Logical Problem of Induction first appeared in French in 1923, and *Geometry in the Sensible World* in 1924. They have been re-issued in French, the former in 1961 and the latter in 1962. They appeared in a single volume in English in 1930, in the International Library of Psychology, Philosophy and Scientific Method, published by Kegan Paul, Trench, Trubner & Co.

The translation in this new edition of *Geometry in the Sensible World* has been done by Mr John Bell, of Christ Church and the Mathematical Institute of Oxford University. He wishes to express gratitude for the very valuable assistance that he received from Mlle Michéle Aquarone, and his gratitude also to Mr Brian Priestley. The translation of *The Logical Problem of Induction* has been done by Mr Michael Woods, Fellow of Brasenose College, Oxford.

The translation has not been easy. While Nicod has natural fluency and a wide command of apt imagery, the French is often subtle and *nuancé*, and to convey the exact shade of meaning in English has sometimes seemed well-nigh impossible. I have gone over the translations word by word, and have had long sessions with the translators, especially in the case of the rather more difficult *Geometry*, in order to attain a consensus with them, which, I believe, has been achieved in all cases.

In 1930 I asked Alfred Whitehead his opinion about the current state of the studies in those fundamental questions on which he had worked in collaboration with Bertrand Russell. He replied that those to whom he had especially looked for further progress among the next generation were Nicod, Ramsey and Scheffer, and added that it was a matter, not only of great grief, but also of concern, to him, that two of these had already died in their youth. Nicod died in 1924, and Ramsey in 1930, a few months before my talk with Whitehead. It has been a pleasure to me during the course of my labours on this book to be able to think what I was doing would have been approved by my two very dear friends, Alfred Whitehead and Frank Ramsey. I did not know Nicod; but I heard him give a lecture at the International Congress of Philosophy in Oxford in 1920.

I am not qualified to assess the bearing of more recent developments in mathematical logic on the analysis supplied in Nicod's *Geometry*. I would, however, hazard the guess that the *relation* between any formal system and the contents of our sensible experience will remain an unsettled question, like so many others in philosophy, for a very long time. I believe that I am qualified to affirm that about the very central logical problems discussed in *Induction* there is still no agreement.

I read these two essays when they first appeared in English in 1930, and was strongly stimulated by them. Indeed I cannot readily think of any philosophical work of the twentieth century which has had so profound an influence upon my mind. This is not the place to repeat what I have already set out in respect of that in the Preface to my *Foundations of Inductive Logic* (1956). But perhaps I may be permitted to say that I believe my book to be one of the very few books since Nicod – mention should, however, be made of Professor Donald Williams – that is based on the kind of premisses that would have been acceptable to him.

I am grateful for the hospitality of the Rockefeller Foundation, which enabled me to carry out part of my work in the lovely Villa Serbelloni on Lake Como.

R. F. HARROD

GEOMETRY IN THE SENSIBLE WORLD

Jean Nicod

TO MY MASTER

MONSIEUR ANDRÉ LALANDE

Professor at the Sorbonne
Member of the Institute

IN TESTIMONY OF GRATITUDE AND RESPECT

Contents

Preface

The premature death of Jean Nicod is more than a source of deep sadness for those who knew him; it means an irreparable loss in the realm of philosophical studies. Besides articles of considerable value, he had completed two theses for the doctoral degree at the University of Paris. The shortest of these is devoted to the logical problem of induction (pp. 193 ff.); the longer, whose text we have here, deals with a problem whose importance has appeared more and more clearly in the last few years: the relation between geometry and sense-perception.

The history of this problem in modern times is well known. Kant asserted that geometry is based on an *a priori* intuition of space and that experience could never contradict it because space constitutes a part of our manner of perceiving the world. Non-Euclidean geometry has led most thinkers to abandon this opinion; although from the logical point of view, it might be easy to maintain that Lobachevsky's work did not conflict with Kant's philosophy. Another stronger but less known argument was employed against Kant; it is the argument derived from the attempt to reduce pure mathematics, at first to arithmetic, and then to logic. The implication was that an *a priori* intuition is no more necessary for abstract geometry than for the doctrine of the syllogism.

However, it was still possible to adopt a point of view which has certain affinities with that of Kant; for example, it was the viewpoint assumed by Henri Poincaré, who maintained that Euclidean geometry is neither true nor false, but that it is simply a convention. In a certain sense, this point of view may still be possible: in all experiment or physical observation, it is *the group* of applicable physical laws which constitutes the object of study, and if the results do not correspond to our expectation, we have a certain choice as to which of these laws should be modified. For example, Henri Poincaré would have maintained that if an astronomical observation seemed to prove that the sum of the angles of a triangle is not exactly equal to two right angles, this phenomenon would be more easily explained by assuming that light does not travel in a straight line than by giving up the system of Euclidean geometry. It is not surprising that Poincaré should have adopted this view; what is surprising, is that the progress of physics should have since shown in its own realm, that this point of view was ill taken. In fact, the eclipse observations undertaken to verify the Einsteinian theory of gravitation are explained usually by admitting both that space is non-Euclidean and that light is not propagated strictly in a straight line. Undoubtedly, it is still *possible* to hold to the view that space is Euclidean, as Dr Whitehead does, but it is at least doubtful whether such a theory furnishes the most convenient explanation of the phenomenon.

In the following pages there will be found a different criticism, more fundamental than the theory of Henri Poincaré. When a logical or mathematical system is applied to the empirical world, we can distinguish, according to Jean Nicod's observation, two kinds of simplicity: simplicity intrinsic to the system and simplicity extrinsic to it. Intrinsic simplicity is the simplicity of the laws that establish the relations among the entities taken as primitive in the system. Extrinsic simplicity is the simplicity of the empirical interpretation of these entities. The points, lines, and planes of geometry give it the character of intrinsic simplicity, because they enable the axioms to be stated briefly; but they do not constitute what is empirically given in the sensible world. Consequently, if our geometry is to be applied to the perceived world, we shall have to define points, straight lines, and planes by means of terms which are at least similar to our sense-data. In fact, this definition is extremely complicated, and thus removes any character of *extrinsic* simplicity from our conventional geometry. To regain this extrinsic simplicity, we must start from data which are not in conformity with ordinary geometry, Euclidean or non-Euclidean; and we must formulate gradually, if we can, suitable logical constructs that enjoy the required properties. We cannot say in advance whether we shall obtain greater extrinsic simplicity by having recourse to straight lines and Euclidean or non-Euclidean planes, although we admit that the possibility of one of the systems implies the possibility of the other and conversely.

Dr Whitehead has examined, from the point of view of mathematical logic, how we can define in terms of empirical data the entities that traditional geometry considers as primitive. His method of 'extensive abstraction' has great value and efficacy in this regard. But this method starts from the knowledge of the completed mathematical system which is the object to be attained, and goes back to entities more analogous to those of sense perception. The method adopted by Nicod follows the inverse order: starting from data of perception, it tries to attain the various geometries that can be built on them. This is a difficult and novel problem. To treat it logically, the author assumes as a starting point an entirely schematic simplicity of sensations, although it is easy to imagine some animals among whom it might exist. In his first example, he shows us an animal possessing only the sense of hearing and a perception of temporal succession, who produces notes of varying pitch as he proceeds up and down the keyboard of a piano. Now, such an animal, if we suppose him endowed with sufficient logical power, will be able to produce two geometries, both, naturally, in one dimension. The animals presented next come nearer to man in their perceptions; although they differ from most of us in that they are logicians and metaphysicians as penetrating as Nicod himself.

The distinction between pure geometry and physical geometry, which has gradually appeared of late, is presented as clearly as possible in Jean Nicod's work, the first part of which deals with pure geometry. This distinction and its consequences are not yet comprehended by philosophers as much as they deserve. In pure geometry we assume as a starting point the existence of a group of entities whose relations have definite logical properties and we deduce from them the propositions of the geometry under consideration. The existence of groups of entities having relations of this nature can in all usual cases be deduced from arithmetic. For example, all the possible triads of real numbers arranged in their natural order form the points of a three-dimensional Euclidean space. The whole question belongs to the realm of pure logic and no longer raises philosophical problems. But in physical geometry, we are confronted with a much more interesting problem because it is far from having been completely solved. We know that experimental physics employs geometry; from this it follows that the geometry which it employs is applicable to the empirical world to the degree in which physics is exact. That is to say, it ought to be possible to find groups of sense-data and relations among these data such that the relations which are derived from these groups may approximately satisfy the axioms of the geometry employed in physics. Or, if the sense-data alone are not sufficient, they ought to be complemented in the same way as they are in physics, by means of inferences and inductions whose use is authorized by ordinary scientific method; for example, the inference which allows us to assume that the moon has another side which we do not see. This point of view is supported and facilitated by the absorption of geometry by physics as a result of the theory of relativity. However, the psychological aspect of this problem has been studied very little, probably because few psychologists possess a sufficient knowledge of modern physics or mathematical logic. We must build a bridge by beginning on both ends at the same time: that is to say, on one side, by bringing together the assumptions of physics and the data of psychology and, on the other, by manipulating the psychological data in such a way that we may build logical constructions that approximately satisfy the axioms of physical geometry. Jean Nicod has, in the last of these tasks, made progress of the highest importance. He has created a method much superior to that of his predecessors. We cannot say yet that the two sides of the bridge meet in his work, but the gap that remains to be filled today is smaller than it was before the writing of the following pages. That is why I recommend the study of them to all those who believe in the value of philosophical research and who are capable of appreciating in this work the rare clarity and beauty of its exposition, which reflects faithfully the equal beauty of the author's life and character.

BERTRAND RUSSELL

1924

Introduction

Experience is the only criterion of the truth of propositions about the bodies around us, or at least of those which are to some extent particular. Let us call the totality of these propositions *physics*, taking the word in its widest possible sense. Physics is an entirely empirical science: it attains certainty or probability in the degree to which it is verified by experience and in this degree only. Its sole claim upon our credence is the exactness with which it tells us what we shall see, hear, and touch, in accordance with what we have seen, heard, and touched. If that is not its only task, it is in any case that by which it wishes to be judged. Everyone thinks it feasible to analyze physics, to the extent that it is verifiable, by reference to the data of sense.

This analysis is still far from being achieved. Writers most concerned with the empirical character of the import of propositions about the material world give us in fact only the most summary indication of what this import is. They take some proposition of physics and say: 'In experience it signifies something like this'. But not exactly this. For if we care to examine the matter more closely, no physical fact has unambiguous expression in a sensible fact. Any sensible fact, we say, can result from a variety of physical causes although they are unequally probable: or again, our senses can deceive us about objects. But on the other hand, these same senses can also un-deceive us by means of sensible facts, which are sometimes far removed and very different from those which we had at first misinterpreted. Thus the manifestation in my experience of an immediate physical fact goes beyond my present observation and extends to the totality of my observations past and future. Because of this, the sensible manifestations of various physical facts are not as distinct from one another as these facts themselves, but, on the contrary, inter-penetrate. If we want the last word about the sensible manifestation of any particular proposition of physics, we must seek it in nothing less than the totality of experience conforming to the totality of physics. With respect to verification, as Duhem has rightly perceived, physics as a whole constitutes a block. It is the form and structure of this block that we wish to investigate, in order to discern in it the sensible content of the laws, simple or complex, ordinary or sophisticated, that constitute our knowledge of nature.

Assuredly this content is already present to our mind. It continually provides us with particular forecasts. But the totality to which it belongs escapes us. It guides us while remaining in the shadow: we know how to make use of it, but not how to apprehend it.

The reason for this strange fact is that the formation and growth of physics are dominated completely by the pursuit of simple laws or, better, of the simple expression of laws. This expression can in fact only be obtained by marking complex things by simple names. For nature is constituted in such a way that it is not the simple things that enjoy simple laws, and so, in order to simplify the laws, we must complicate the meanings of the terms. Energy, matter, object, place, and time themselves, taken in the physical sense, and, more generally all the terms which are employed by physics and which do not express elementary observations, derive all their utility, their whole *raison d'etre*, from this compelling need for simple and striking statements of the laws of the sensible world.

The real complexity of these laws is thus hidden within the very meaning of the new terms. It stands out naturally in their application; but it ceases to harass the intellect. It even ceases to be distinctly perceived by it. The intellect embraces, one might say, passionately, these new terms which manifest such an agreeable order. The sensible world is thus eclipsed by its image; and we still have to learn to discern it in the physics that portrays it.

Up to now this investigation has not been performed, and as a result we believe in laws which are founded only on experience without perceiving exactly what they mean in terms of experience. True, it is a difficult task and, still more, a long one. But above all it did not figure in the plans put forward by the philosophers. For they only concerned themselves with the sensible content of judgments about reality in order to extract from it an argument on the general nature of the matter of physics. This argument sprang from the mere existence of this sensible content, not from its more particular constitution; and since this existence is so little in doubt that the most summary indication is sufficient to make it obvious, they stopped at that. The empirical analysis of nature, from the instant that its feasibility was apparent, no longer seemed worth performing.

But *we* must conclude otherwise. Even if it should not prove useful in his metaphysics, what philosopher would not be interested in recognizing the sensible order around us, so evident and yet so indistinct, which forms the fabric of our lives and of our science? This is the target at which we are aiming. We hope to approach it by investigating the sensible aspect of geometry. It is in fact impossible to arrive at an exact conception of the order of our sensations if we are encumbered with some false or confused notion of space.

This study could be an introduction to the analysis of physics in terms of experience. It is also the first step. For we shall find that the universal order of space to which each statement of physics apparently refers is in truth none other than the very totality of physical laws. The properties of space already constitute the most general schematism for physics and nothing else. Thus – we shall convince ourselves of this as we proceed – the study of the spatial network of a sensible universe is the study of the form and interrelations of all its laws.

In this work we intend to investigate the ways in which geometry assists physics; how its propositions apply to the order of the sensible world; how our knowledge of them serves in the description of experiences and laws. For every statement of physics is shot through with geometry: every prediction of a sensible fact involves a certain arrangement of objects and observers, expressible in geometric terms.

We ask how geometry is thus manifested in nature: not why this is so. We seek what the fact consists of, and not the reasons which render it possible or necessary. Indeed, analysis should precede explanation, for analysis is always possible, whereas explanation is not.

In this problem, geometry is presented as a form for which the sensible world provides the matter. The natural order is to investigate the form, then the matter, and finally the particular way in which one is manifested in the other. First of all, therefore, let us familiarize ourselves with geometry as a formal and entirely abstract science of the consequences of certain principles involving terms and relations whose interpretations are not specified. Then let us determine the terms and relations we apprehend in sensible nature. Finally, let us try to discover the interpretations, constructed from these terms and relations, which conform to the terms and relations of geometry and manifest its laws in experience.

Throughout the course of this work I have been greatly assisted by the advice and benevolent criticism of M. A. Lalande. I express here my warmest gratitude to him.

I am also indebted to M. E. Cartan for several valuable observations.

Part One

GEOMETRIC ORDER

CHAPTER I

Pure Geometry is a Logical Exercise

What is geometry, regarded as a pure form? It is what one can know of it while being ignorant of its object; it is what one can understand in a treatise on geometry without knowledge of the nature of the entities it discusses.

In Kant's time, this point of view had still not been formulated. For geometry, which since Euclid had been tending to free its demonstrations from the material furnished by figures, in order that they might be based solely on reason, had not yet succeeded in doing so. Its proofs, when deprived of the figures that illustrated them, seemed without force; the very connections between the propositions appeared to derive from their subject matter, and not from the purely logical relations between them. Consequently, all geometrical knowledge was conceived as inseparable from the apprehension of an elemental substance, namely space, which, having communicated its order to the sensible world, played, on the other hand, in regard to that world the role of a form. Thus the still imperfect character, the still indecisive nature of the geometers' demonstrations, appeared to the philosophers as a special mystery, and committed them to cumbrous theories designed to account for the alleged existence of proofs not deriving their force from common logic.

However, the present perfection of geometrical science allows us at last to conceive the problem in simple terms. In fact, while the philosophers were speculating on the extra-rational character of geometrical proof, the geometers succeeded in doing away with this character. They laid down the principle that a proof by figures is but a sketch of a proof. In the necessity of an appeal to intuition they saw evidence of the presence of a lacuna, an indication of the use of an implicit principle they were striving to formulate. Only after it had been presented as a wholly discursive sequence would they regard a proof as formally correct.

In order to grasp a proof in such a formally perfect state, it is no longer *necessary* to illustrate it with a figure, to relate it to a substance, or to assign definite significations to the geometrical expressions it brings into play, but whose meanings have no effect on its force. It is *possible* to convince oneself that the theorems are derived from the axioms and postulates without knowing what the terms *point*, *straight line* and *distance* signify: no present-day geometer would deny this. In the process of becoming rigorous, that is to say, explicit, geometric proof has also become independent of any object.

This is by no means a paradoxical development. On the contrary, it marks the end of the paradox involving the opposition between geometrical reasoning and all other forms of reasoning. For the validity of sound reasoning, expressed without implied assumptions, derives from its form alone: it is quite independent of the truth and even of the meaning of the propositions it links together. This important fact may cause surprise, but it cannot be doubted. By freeing itself from figures and ignoring the meaning of the material terms it employs, geometric reasoning has simply re-entered the realm of common reasoning.

It is therefore possible today – as was not the case a century ago – to adopt a first view of geometric science by maintaining complete abstraction fiom any object. It is thus presented as a sequence of formal, and in a certain sense blind, reasonings, which draw their consequences from a group of premises formulated in terms of entities whose nature, being independent of the arguments, remains completely indeterminate. This is the universal aspect of geometry. It is this form, still devoid of all reality, that it is expedient for us to fix in our minds. For, by conceiving it as independent of any object in the first place, we are preparing ourselves to determine, without preconceived notions, to which objects in the universe it can actually be applied.

Let us then assume that we have not been taught geometry at school and that we are unacquainted with any of the specific terms of this science. Doubtless the things with which this science is commonly acknowledged to deal are familiar to us. But we may suppose that we have not been taught their technical names and that, like the young Pascal, we call a line a 'bar' and a circle a 'round'. Now imagine that one of those books on geometry which aim at rigour alone and scorn figures is placed in our hands. What shall we make of it? Let us attempt to read it in any case.

It is constructed from a small number of initial statements called 'axioms' or 'postulates' and of others called 'theorems' which appear to derive from the former by means of passages called 'demonstrations'. Now, although we understand the expressions of current parlance, and in particular the standard logical expressions, we are nonetheless unfamiliar with all the specifically geometrical terms, for example, 'point', 'line', and 'distance'. At first these new terms strike us as being extremely numerous. However, we soon notice that they are for the most part introduced as simple abbreviations for complex expressions, in which only a few unknown terms appear; and these are invariably the same ones, namely, those which figure in the initial propositions. Typical examples will be the class of 'points', the relation of three points 'in a straight line', and the relation between two pairs of points 'separated by the same distance'. Thus the term 'sphere' will appear as an abbreviation for the complex expression 'class of points separated from a given point by a constant distance'.

We have made an inventory of the unknown terms, and reduced their number to three. We have not eliminated them, however. Since we have no knowledge at all about the meanings of those that remain, we are forced to admit that we are equally unaware of the meanings of the 'axioms', the 'postulates', and the 'theorems' in which these remaining terms appear. But to our surprise we understand the intermediate steps – the 'demonstrations' – perfectly. In spite of the fact that they too contain instances of these unknown terms that puzzle us, we need only understand the surrounding words of everyday discourse, all of which are part of the logical structure of language, for us to be in a position to follow the argument step by step, to grasp its development, to appreciate its ingenuity and to recognize its exactness.

There is something surprising in the fact that the rigour of a demonstration may be recognized without any knowledge whatsoever of its subject matter. We are astonished when we discover our ability to proceed, so to speak, with our eyes shut. But this characteristic strength of form as such re-appears in the simplest of reasoning, which is valid in a given case only on condition of its being so on the same basis for all cases, possible and impossible alike. It is this fact that constitutes the nature of logic, remarkable but none the less universal.

So, what do we learn from reading this book? We may express it as follows: 'I do not know what the author of this book means by a "point", and still less what he means by three points in a "straight line", or two pairs of points separated by the same "distance". But I do know that if these three things really have, as he asserts, the properties stated in the axioms and postulates, then they must inevitably have the properties stated in the "theorems".'

The fact that we have been able to construct a chain linking the propositions in which these three terms figure *without having to assign meanings to them*, enables us to rise to a more general viewpoint. Instead of granting these terms definite, but unknown, meanings, we may regard them as variables, simple instruments enabling us to express the following universal truth. 'Let a class π, a relation R having as terms three members of π, and a relation S having as terms two pairs of members of π, satisfy the axioms and postulates – in other words, suppose that the assignment of the three "meanings" π, R, S, to the three expressions *point*, *in a straight line*, and *separated by the same distance*, transforms the axioms and postulates into *true* assertions. Then under these conditions, π, R, S, also satisfy the theorems.'

A geometrical proposition therefore ceases to be a determinate proposition, capable of being true or false on its own merits. It is no more than a blank propositional formula capable of engendering a whole host of different propositions, some true, some false, according to the meanings ascribed to its variable terms. In other words, it is merely a *propositional function*. All we can learn in our state of ignorance from the geometry book that has fallen into our hands is that the propositional functions called axioms and postulates [1] imply (whatever meanings are assigned to the expressions) the propositional functions called theorems.

Let us now close the book and ask ourselves why the author wrote it. It may have been only the attraction of an adventure in logic, or the pleasure of deducing the implications of a group of statements chosen, like the rules of a game which appeal to the mind for the

(1) The difference between a postulate and an axiom, which is merely that one is less self-evident than the other, does not exist for us because both are devoid of fixed meaning and consequently have no degree of self-evidence whatever. In order to simplify the language, we shall henceforth refer to all the premises of a geometry book as *axioms*. This moreover is the usage of several modern geometers.

diversity and harmony of their consequences. On the other hand, the author may have composed in accordance with nature. Has he not modelled his axioms on the properties, demonstrated or conjectural, of certain entities which may be found in his universe and perhaps also in ours? So let us try to discover or at least conceive, one or several systems of meanings satisfying our author's axioms: we shall call such a system of meanings a *solution of this group of axioms.*

The domain of numbers furnishes the first example. We ascribe the following meanings:

(1) to the variable class of *points*, *the class of ordered triads of real numbers with attention paid to their signs*;

(2) to the variable relation *in a straight line*, *the relation* among three triads of real numbers (x_1, y_1, z_1), (x_2, y_2, z_2), (x_3, y_3, z_3) expressed by the equations

$$\frac{x_1 - x_2}{y_1 - y_2} = \frac{x_2 - x_3}{y_2 - y_3}, \frac{x_1 - x_2}{z_1 - z_2} = \frac{x_2 - x_3}{z_2 - z_3}$$

(3) to the variable relation *separated by the same distance*, *the relation between two pairs of triads of real numbers* $(x_1, y_1, z_1), (x_2, y_2, z_2)$; $(x_3, y_3, z_3), (x_4, y_4, z_4)$ expressed by the equation

$$(x_1 - x_2)^2 + (y_1 - y_2)^2 + (z_1 - z_2)^2 = (x_3 - x_4)^2 + (y_3 - y_4)^2 + (z_3 - z_4)^2.$$

We know that the system of 'meanings' (1), (2), (3) satisfies the axioms of Euclidean geometry.

This geometry, therefore, admits a purely arithmetical interpretation or 'solution'. However because of the fact that abstract geometry is very often confused with its application to a particular interpretation known as 'space', the arithmetic interpretation is generally introduced indirectly as a measure of space. But this detour is superfluous. As soon as geometry is conceived as an essentially meaningless schema standing outside all application, we can see that it applies *directly* to numbers, and that we need not regard them as measures or representations of a definite substance. Thus, by choosing the purely arithmetical notions (1), (2) and (3) as interpretations for *points*, *rectilinearity* and *congruence*, we see that the axioms and, in consequence, the theorems, are resolved into arithmetical propositions. For example, the axiom which says that if the points a, b, c are in a straight line, and if the same holds for the points b, c, d, then the points a, b, d are likewise in a straight line, under this interpretation becomes: If the numbers x_a, y_a, z_a; x_b, y_b, z_b; x_c, y_c, z_c satisfy

$$\frac{x_a - x_b}{y_a - y_b} = \frac{x_b - x_c}{y_b - y_c}, \frac{x_a - x_b}{z_a - z_b} = \frac{x_b - x_c}{z_b - z_c}$$

and if the numbers x_b, y_b, z_b; x_c, y_c, z_c; x_d, y_d, z_d satisfy

$$\frac{x_b - x_c}{y_b - y_c} = \frac{x_c - x_d}{y_c - y_d}, \frac{x_b - x_c}{z_b - z_c} = \frac{x_c - x_d}{z_c - z_d}$$

then the numbers x_a, y_a, z_a; x_b, y_b, z_b; x_d, y_d, z_d satisfy

$$\frac{x_a - x_b}{y_a - y_b} = \frac{x_b - x_d}{y_b - y_d}, \frac{x_a - x_b}{z_a - z_b} = \frac{x_b - x_d}{z_b - z_d}$$

and this is certainly a theorem of arithmetic.

It is true that the arithmetical interpretations of the primitive expressions which appear in our geometry book are not *simple* and seem artificial because of their complexity. Is it not strange that a point can be a class of three numbers? Not at all, because 'point' is merely an empty word before this interpretation presents itself.

The discovery of the first system of meanings satisfying a group of axioms is always of great logical importance, for it proves that these axioms do not contradict each other, and this is the only known proof. Arithmetic guarantees the consistency of the axioms of geometry through this discovery. But there may be 'solutions' of this group of axioms outside the domain of numbers: the investigation of this possibility forms the purpose of this work.

CHAPTER II

The Formal Relationship Between Various Geometric Systems

We could end this introduction here without taking note of the fact that geometry can be put into more than one form, for the reason that all these forms are equivalent. However, we shall adopt a more general viewpoint by taking their plurality into account and by seeking the precise nature of their equivalence.

The geometry that we learn at school is the study of the consequences of a group of axioms which can be formulated in terms of three fundamental expressions: point, straight line, and distance. It is customary to regard this group of axioms as the fixed and necessary foundation of euclidean geometry. But it is nothing of the sort, indeed it is only one of its possible foundations, because this geometry, like a polyhedron, can rest on a multitude of different bases.

Even before abandoning the primitive classical notions in favour of others, it is possible to replace certain axioms by other propositions which formerly held the rank of theorems, and which, inversely, enable us to demonstrate the old axioms which they replace. In this way we can formulate several equivalent lists of axioms for the euclidean geometry of point, line and distance. It is true that some of these lists are shorter than others, and that one of them can prevail in virtue of its brevity and elegance; but they are all equally correct. Thus, for a given system of primitive notions, the order of the statements of euclidean geometry and their distribution into axioms and theorems are by no means immutable. But further, we can modify the primitive notions: this gives rise to inversions of a more radical nature.

From the current system of point, straight line and distance it is easy to eliminate the straight line as a primitive expression by defining it as the class of points equidistant from three given points. If the geometer does not normally proceed in this manner, it is because he esteems progression more than he does economy, and because he wishes to emphasize the distinction between projective and metric properties. But when we consider geometry as a whole, it is often convenient to be able to regard it as the science of the sole relation of the congruence of pairs of points.

Is this relation, then, its irreducible foundation? No, because it can be replaced by many others.

In spite of the fact that straight line and distance alone appear in his explicit axioms, Euclid occasionally has recourse to the properties of the totally different notion of *displacement*; because he never took the trouble to codify the use he made of it, people have wished to take this gap in his demonstrations as proof of the existence of a certain amount of unformalizable intuition. However, the present state of knowledge should lead us to adopt a quite different point of view. On the one hand, the geometry of straight line and distance, supplied as it is with a fuller list of axioms, no longer makes use of the support provided by the extraneous notion of the displacement of figures. On the other hand, this notion may be taken as a sole foundation for geometry. And so it happens that alongside the geometry of straight line and distance, finally made conformable to the ideal of rigour that Euclid sought, a geometry entirely founded on the idea of displacement has emerged from the matter that vitiated some of the master's proofs.

But it is not only congruence or displacement which can serve as a sole primitive relation. A multitude of other relations could be adopted, for example the relation of five points on a sphere: it is indeed possible to define all geometrical notions in terms of the sole relation of 'sphericity'; this is only an example taken at random. Geometry may be presented as the science of the logical properties of any one of these various point relations. The various deductive systems obtained in this way would doubtless differ in overall elegance, but there would perhaps not be one which did not have a particular advantage over the others.

When we pass from one of these equivalent geometries to another, each of which is formulated in terms of various primitive expressions, the order of the propositions, and in particular their distribution into axioms and theorems, is naturally modified. The geometry of congruence arises from the properties of congruence, that of the sphere from the properties of the sphere, that of displacement from the properties of displacement, and the order in which we encounter the propositions varies with the point of departure.

But is it only the order that varies? These propositions which appear to be the same in all systems, when the hierarchy alone is changed, are they truly identical? Let us examine this closely: their identity is only apparent. Indeed, the geometry of congruence defines all geometrical entities in terms of congruence, the geometry of the sphere defines them in terms of spheres, the geometry of displacement defines them in terms of displacement: in each of these three geometries, the word straight line (for example) is an abbreviation of a different expression – a function of congruences, a function of spheres, a function of displacements. Thus the same statement does not have the same meaning throughout. The proposition: *Any two points are part of the same straight line* signifies in the first system: *For each pair of points*, x, y, *there exist three points*, a, b, c, *such that* x and y *are both equidistant from* a, b, c. In the second system, this same statement means: *Each pair of points* x, y, *belongs to two distinct classes of points* M, N, *neither of which forms part of a sphere with three given points*. Finally, in the third system it has the following meaning: *For each pair of points* x, y, *there is a displacement which maps* x *on to itself and* y *on to itself*. At bottom each system knows its own primitive expressions only, and is incapable of discussing anything else. If we succeed in finding the same statements here and there, that is thanks to the device of identical abbreviations which conceal different expressions, and this is merely a clever play on words.

But in view of the fact that, once the defined has been replaced by its definition, the various systems which result from the change in primitive expressions no longer contain anything but distinct propositions, in what sense can they be regarded as so many facets or aspects of one and the same geometry?

Let us not reply too hastily: 'through the identity of their subject matter'. For this identity is not literal identity, and needs to be defined. Besides it arises from a purely formal relationship: merely by comparing two systems of euclidean axioms, without assigning a meaning to their primitive expressions, we should be able to convince ourselves that they necessarily apply to the same worlds. In order to grasp the precise nature of this identity of subject matter, and also to understand its justification, it is advisable to clarify the formal relationship which ensures its existence.

Let us compare the two systems which present euclidean geometry as the science of an undefined relation known as *congruence* of two pairs of points and, as the study of an undefined relation called *sphericity* of five points. The first speaks only of congruences, but presents sphericity as a certain function of congruences. Likewise, the second speaks only of sphericities, but it employs the term congruence as an abbreviation for a certain function of sphericities.

These two functions are the following:

In the geometry of congruence, *sphericity* is the name given to the relation holding among five points x_1, x_2, x_3, x_4, x_5 which holds when there exists a point y such that the pairs x_1y, x_2y, x_3y, x_4y, x_5y are congruent.

In the geometry of sphericity, *sphere* is the name given to the class of points x in the relation of sphericity with four given points, when this class exists. When this class does not exist, we shall say that any one of the four given points is situated in the *plane* of the three remaining ones. If the two points x, y are the only common points of a certain sphere with the planes M and N respectively, and if z is a point common to M and N, we shall say that z is *equidistant* from x and y. Finally we shall say that the pairs of points ab, cd are *congruent* whenever there are points $x_1, x_2, \ldots, x_n$ such that each interior member of the sequence $a, b, x_1, x_2, \ldots, x_n, c, d$ is equidistant from its two neighbours.

The two terms, congruence and sphericity, are thus found in both systems. But their meanings cannot be the same, because their relationship through composition is reversed [by passing from one system to the other]. Each in its own system is simpler than the other. In one system, sphericity denotes a construct from congruence; in the other, on the contrary, congruence denotes a construct from sphericity: we cannot at the same time admit both possibilities if the two terms have the same meaning. In order to extract the true connection between the two systems, we must delve beneath this identity of words. We must recognize the artifice, and attempt to discover the reason behind its usefulness.

We already know the reason: the application in different systems of the same words to things whose relationships through composition are inverse, gives rise to complete identity of the *propositions* stated in all the systems. It is true that this identity is, like the first, merely verbal. But it makes manifest a certain formal correspondence. It shows that it is *possible* to 'translate' the simple expressions of a system into compound expressions of another in such a way that the axioms of the first system (and consequently all its propositions) become translated into propositions of the second system. This is the relationship which logically binds together all conceivable systems of euclidean geometry. It enables us henceforth to define this geometry in a general way. Starting from any one of its forms, for example the science of the congruence of pairs of points, we embrace in thought all possible forms of Euclid's geometry through the concept of systems which are 'translatable' into this one (and conversely).

This relationship is familiar to the algebraist. It is none other than the transformability of one function into another by a change of variable. Just as the function $y^2 + 2xy + x^2 + x + y$ is transformed into the function $u^2 + u$ by setting $u = x + y$, the function: *If there is a point* y *such that* x_1y, x_2y, x_3y, x_4y, x_5y, *are congruent*, *and a point* z *such that* x_2z, x_3z, x_4z, x_5z, x_6z *are congruent*, *then there is a point* u *such that the pairs* x_1u, x_2u, x_3u, x_4u, x_6u *are congruent*, is transformed into the function: *If the points* x_1, x_2, x_3, x_4 ,x_5, *and the points* x_2, x_3, x_4, x_5, x_6 are in the relation of sphericity, then so are the points x_1, x_2, x_3, x_4, x_6 if we write *sphericity = relation holding among five points such that there is a sixth point forming five congruent pairs with them.* We need only to extend the formal relationship in question from numerical functions of numerical variables to any functions of any variables whatever.

Suppose in particular we are given two propositional functions $F_1(x_1, \ldots, \alpha_1, \ldots, R_1, \ldots)$ $F_2(x_2, \ldots, \alpha_2, \ldots, R_2, \ldots)$ in which the indeterminates can be individuals $x_1, x_2, \ldots$, classes $\alpha_1, \alpha_2, \ldots$, and relations $R_1, R_2, \ldots$. The number of terms appearing in the two functions and their logical types need not necessarily be the same; however, let us recall that a group of propositional functions can always be replaced by a unique function – their logical product.

We establish a correspondence between each of the indeterminates of F_2 and a certain *logical* function of the indeterminates of F_1, that is, an expression containing only logical terms apart from these indeterminates, by setting

$$x'_2 = f\ (x_1, \ldots, \alpha_1, \ldots, R_1, \ldots), \text{ etc.}$$
$$\alpha'_2 = g\ (x_1, \ldots, \alpha_1, \ldots, R_1, \ldots), \text{ etc.}$$
$$R'_2 = h\ (x_1, \ldots, \alpha_1, \ldots, R_1, \ldots), \text{ etc.}$$

Now replace in F_2 all the expressions f, g, h, by the simple terms $x'_2, \ldots, \alpha'_2, \ldots, R'_2, \ldots$. If we find that we have

$$F_1\ (x_1, \ldots, \alpha_1, \ldots, R_1, \ldots) = F_2\ (x'_2, \ldots, \alpha'_2, \ldots, R'_2, \ldots),$$

we shall say that F_1 is *transformable* into F_2.

If we now denote by F'_1, F''_1, ... the *logical consequences* of F_1, that is the theorems which result from the set of axioms F_1 and which naturally are functions of the same indeterminates, it may happen that it is no longer the function F_1 itself which is transformable into F_2, but one of the functions F'_1, F''_2, ..., which derive from it [F_1], (a single letter being used to represent the logical product of several theorems, as before that of several axioms). To indicate this more general formal relationship of which the first is a special case, we shall say that the function F_1 *contains* the function F_2. Finally, when the functions F_1 and F_2 contain each other, we shall say that they are *inseparable*. This is the definition of the general formal relationship between two systems of equivalent principles, whether or not expressed in terms of the same primitive expressions: this relationship *constitutes* their equivalence.

CHAPTER III

Material Consequences of this Relationship

Our problem is one of matter and not of form. We propose to search for the objects which satisfy the axioms of euclidean geometry. But these axioms are not unique: we have the embarrassment of a choice between a multitude of possible systems. The problem thus breaks down into a whole host of problems, each of which is concerned with the objects capable of satisfying a particular system of axioms. But the formal connection between these systems re-establishes their unity, since it provides an *a priori* relationship between the subject matters to which they apply.

Suppose we have two inseparable systems of principles $F(X)$, $\Phi(Y)$, formulated in terms of two primitive expressions X and Y. If these expressions were more numerous, the reasoning would still remain the same.

In general, the expressions X and Y do not admit a common meaning. This incompatibility of possible solutions of $F(X)$ and $\Phi(Y)$ is manifest in the case in which X and Y are of different logical types, where X for example is congruence, a relation between two pairs of terms, and Y is sphericity, a relation among five terms. *Two inseparable systems of axioms are not in general satisfied by the same interpretations:* they are, more often than not, *incompatible.*

But each solution of the one logically furnishes a solution of the other. Indeed, because $F(X)$ contains $\Phi(Y)$, there is a logical function

$$Y' = f(X)$$

such that $G(X) \equiv \Phi(Y)$ is a consequence of $F(X)$. Now suppose that a certain interpretation X_1 satisfies $F(X)$, in other words, we have $F(X_1)$. Construct the interpretation

$$Y'_1 = f(X_1)$$

This second interpretation, formulated logically from the first satisfies Φ in view of the fact that f is a logical function. For $\Phi(Y'_1)$ arises from $F(X_1)$, and $F(X_1)$ is true.

Thus the two equations $x^3 + 3x^2 + 4x + 3 = 0$ and $y^3 + y + 1 = 0$, which can be transformed into one another by setting $y = x + 1$, have no common solution, but nevertheless each solution of the one provides a solution of the other. Likewise, no interpretation can possibly satisfy both the axioms of congruence, say $F(X)$, and those of sphericity, say $\Phi(Y)$ at the same time, but each interpretation X_1, satisfying the axioms of congruence, supplies an interpretation Y'_1 satisfying the axioms of sphericity: and this interpretation is the one which the definition $Y' = f(X)$ of sphericity in the system of congruence assumes when interpretation X is assigned to congruence.

From a 'solution' of a system of axioms, we can thus logically construct the 'solutions' of all systems inseparable from it. And *logically*, that is to say without introducing any subject matter. All these conjugate values have the same elements of meaning. They differ only as do addition and subtraction. They are different concepts based on the same facts; their differing laws express the same state of affairs.

We can, therefore, say that two inseparable systems of axioms are true of the same realities. But let us remember that this expression is still undefined, that what we call the same reality provides the raw material for a multitude of logically different entities, whose different laws nevertheless betoken the same order. Were this not the case, we would be confronted with a clash between the identity of the realities in which inseparable systems of axioms hold and the mutual incompatibility of their solutions. We would think that these systems which accompany and yet exclude each other can lay no claim to truth: they are no more than artifices. But this is rather muddled thinking, because these systems are applicable to the same things but not to the same specific entities. Before the assignment of meanings to its primitive expressions, a system of axioms is neither true nor false, since it is but an empty form. On the other hand as soon as we settle these meanings (and not merely the domain which provides the raw material), they can satisfy one system only. The geometries of congruence and sphericity both hold in the same domain, that of numbers, but they are not true of the same entities in this domain: a given arithmetical relation either possesses or does not possess the properties of congruence, or the properties of sphericity;

none can have both simultaneously. The existence of formal systems which are inseparable and incompatible is a commonplace. It does not in the slightest diminish the truth or falsehood of the applications of these systems.

Henri Poincaré sometimes seems to lose sight of this. In fact he points out that non-euclidean systems can be translated into standard systems, and concludes that the question of the validity of the one or the other type is devoid of meaning, or at all events has only a restricted meaning. Now if non-euclidean geometries are translatable into any one of the euclidean geometries and if, as we can easily convince ourselves, the reverse translation is also feasible, it follows from our preceding remarks that non-euclidean geometries differ from their euclidean counterparts in the same way as these latter differ from each other, except for the preservation of the same words 'line' and 'congruence' as primitive expressions for different properties. Thus the non-euclidean systems belong to the set of systems obtained from a given euclidean system by a simple change in primitive expressions. This general collection shows remarkable unity, but does not justify Poincaré's conclusion. Doubtless, if we simply consider a fixed domain, such as the number system, or the physical world, all these geometrical systems hold at the same time, *because their interpretations remain indeterminate*. But tell us what arithmetic, or physical, relations you call congruence and rectilinearity: the problem of determining whether these relations behave in a euclidean or non-euclidean manner could not be clearer.

It is no less true that, in any domain, if there exist terms and relations satisfying the axioms of a given geometry, then *others* can be found to satisfy any system of geometric axioms whatsoever. We thus remain free to choose the terms and relations in such a way that the structure of the domain is expressed by the system we select. But certain among all these groups of axioms are simplest in essence: do they not from this fact derive an advantage over the others, an inalienable advantage since it derives from an intrinsic quality?

Poincaré thinks so; this explains why he does not realize that a change in the universe or in our knowledge of it can deprive traditional geometry of its privilege of simplicity. However, this viewpoint takes into account only of one type of simplicity and complexity, while there are in fact two.

On the one hand a system of axioms can be essentially simpler than another: it may contain fewer, or shorter, axioms, or it may have a smaller number of terms in all, or contain fewer primitive expressions. This type of simplicity has no connection with the meanings which these primitive expressions are capable of assuming; it is related to the form of the system and we may call it *intrinsic* simplicity. It is the only type that Poincaré considers.

But we must also consider the simplicity of the *meanings* which may be assigned to the primitive expressions of the system. What relations can be taken as interpretations of *congruence*? The geometry of congruence does not tell us: we must find out for ourselves. Similarly, the geometry of sphericity does not indicate the meanings its fundamental relation is capable of assuming. From each suitable interpretation of congruence we can construct a suitable interpretation of sphericity, and from every suitable interpretation of sphericity we can construct a suitable interpretation of congruence. But in a given domain, which of these two expressions admits the simpler interpretation? This is what would establish an advantage of one over the other in the domain considered; and this is exactly what we know nothing about *a priori*. For this second type of relative simplicity is *extrinsic*. It may be inverted through a change in the domain of interpretation. It is in no way related to the form of the systems before us.

Indeed, from the simplicity of the formal laws governing certain expressions we *cannot* infer the simplicity of the meanings these expressions are capable of assuming. The fact is that these two types of simplicity are mutually independent. In order to simplify the laws, it is often necessary to complicate the entities they link together; the expert knows this only too well. It is true that we find this divergence displeasing, and that we tend to expunge it from our vision of the world: we then show that complex interpretations which obey simple rules in fact conceal simple objects we know nothing about. Such is energy, such is matter, such is interval in the theory of relativity, and such is in general each entity which enjoys the ideally simple property of invariance. But we are not deceived by this metaphysic correcting the world and forcibly transmitting the simplicity of their laws to the notions themselves. We know perfectly well that the applicable interpretations of entities of this sort have in general an irreducible constitutional character.

So far as geometry is concerned, is it not clear that experience alone distinguishes, among the multitude of groups of possible primitive notions, the one whose interpretation in nature is simplest? If light rays travelled in circles, a simple glance would enable us to determine no longer as is customary the relation among objects in a straight line, but the relation among objects arranged along the same circle, and this 'circularity' would then have a simpler empirical meaning than rectilinearity. If light rays assumed the form of Lobatchevsky's straight lines, then Lobatchevskian geometry, although perhaps more complex than that of Euclid, would nevertheless be applicable to more elementary natural entities. No matter how complicated or disadvantageous, a geometrical system always possesses a corresponding universe. In this universe it applies to simpler things than do any of the systems which are simpler in themselves than it is; and the order of nature cannot be given more elegant expression except by formulating it in terms of more complex, and thereby more artificial, conceptions. That one geometry rather than another is appropriate for a given universe, or to some of its domains, is certainly a question of simplicity. But it does not depend only on the formal simplicity of the system under consideration; it is also dependent upon the simplicity of the interpretation offered it. Is there a form of geometry that is best for physics in all its branches, and what is this form if it exists? This is a positive empirical question. It cannot, therefore, be resolved with certainty. But the probability of one solution or another depends on the state of physics and can vary with it.

If we might be allowed to indicate what appears to be the flaw in Poincaré's discussion, so sound in other respects, it is that, in his struggle against the idea of a space independent of things, he does not quite make a clear break. He shows that space in itself has no properties, that it is unapprehensible, that it is nothing; but instead of rejecting this intermediate nonentity and then applying geometry directly to real objects, he invariably reverts to conceiving it as the science of space. This is why geometry does not seem true to him without introducing conventions, since any structure can be predicated of space without any consequences apart from a compensating change in the expression of the laws of the world: and this is but an extremely tortuous way of asserting that these laws, as soon as they provide a world satisfying a given geometry, can equally well assume the form of any other geometry, if we suitably reverse the physical concepts in terms of which these laws are expressed. But the fact that particular physical interpretations satisfy the axioms of a particular system does not involve a convention; on the contrary, it is the degree of simplicity of these interpretations and not the intrinsic simplicity of the system, which provides a measure of the fitness of this system for this application.

A more complex geometrical system may thus be more suitable than another in a given situation. It cannot be said that the mathematically superior systems admit the simplest interpretations in nature. For the pure geometer, the relation among the points of a straight line is simpler than that among the points of an ellipse, because its laws are simpler: but if we suppose that nature presents a geometric structure, which of these two relations is illustrated more simply? We have no inkling, and indeed the formal advantage of the first relation over the second does not seem to be the ground for any probability in its favour.

CHAPTER IV

Points and Volumes

The time has come for us to point out that all the geometric systems of which we have spoken up to now contain an expression in common. Although the primitive relations may vary, the primitive terms remain in all cases the same: these are the *points*. Could the point be an indispensable element? Now nature does not in fact provide simple terms possessing the properties that the geometer ascribes to points. In order to obtain such terms, it seems necessary to posit, over and above the apprehensible terms, others which differ from these in their very essence. This is how we usually qualify the application of geometry to the sensible world, since we suppose each complete and rigorous geometry to be founded on points which are 'simple and indivisible'.

We already know that such a presupposition is not justified. Thus even if it is true that every geometry requires points as ultimate terms, it does not follow that the term 'point' is capable of assuming a simple, as opposed to a complex, meaning. We have proof of this in the arithmetical interpretation in which a point is a class of three numbers. Geometry could also adapt itself to the absence in nature of a simple meaning for the term 'point', provided some complex physical concept were to assume its role.

But geometry itself puts us on the track of such a concept. For it is untrue that it necessarily regards 'point' as a simple term. We can conceive systems which introduce point as a composite term, composed of terms more easily interpretable in nature.

Why is the existence of such systems not better known? It is because they are of recent discovery; and also because of their lack of interest to the pure mathematician in view of their considerable intrinsic complexity. They are of interest to us because they complete the demonstration of our thesis by presenting the last remaining expression which might seem to be necessarily primitive in every rigorous formulation of geometry, 'point', as complex and definable in its turn. Nevertheless, these formulations constitute that aspect of geometry which turns out to be closest to nature. This is a suitable point to discuss them briefly.

Instead of speaking of *points* and *relations among points*, these geometries refer to *volumes* and *relations among volumes*; just as elsewhere we say that a certain class of points is a volume, here, conversely, we say that a certain class of volumes is a point. And nature without doubt presents us with volumes rather than points.

The idea of going from volume to surface, and thence to line and point is by no means new. There is something natural about this notion, and many geometries devote a few preliminary remarks to it. In fact we often say that a surface is the limit of a volume, that a line is the intersection of two surfaces and a point is the intersection of two lines. However, this does not define them. For what kind of object is the limit of a volume? It is not a volume; it must therefore be a new kind of object whose existence is merely postulated. Similarly, what is the intersection of two surfaces or two lines, if not some new entity, suggested but not explained by these words? This is not the way we frame definitions in a rigorous geometry. A mathematical definition must be a synthesis: it must mould the content of the new expression from the old terms and relations alone. In the geometry of points, a volume is a class of points interrelated in certain ways, and relations between volumes derive from relations between their points. Similarly, in a geometry which is based on the concept of volume alone, a point must be a class of volumes interrelated in certain ways, and relations among points can only be relations among their respective volumes. Does such a geometry exist? It occurs, for instance, in the work of Mr A. N. Whitehead.(1)

Mr Whitehead assumes a point of view which differs from ours. He proceeds from an analysis of the terms and relations which nature supplies, and seeks a combination of these objects which has the properties of a geometric point. But this combination is itself a part of geometry, for the natural objects it combines must already possess the properties of certain geometric entities, namely, the *volumes*. This explains why Mr Whitehead's construction, in its capacity as an inquiry into pure geometry, is to all intents and purposes an

(1) B. Russell, *Our Knowledge of the External World*, Chap. IV. Cf. A. N. Whitehead, *Principles of Natural Knowledge* and *The Concept of Nature*, where the method of 'abstractive classes' is applied to the four-dimensional world of the theory of relativity.

investigation into a geometry of volume. It seems interesting to summarize it in this form.

We begin by recalling the definition of volume in point geometry. Consider a set of points A, and a point p. If each sphere with centre p contains in its interior at least one point of A,[(1)] we say that p is a *limit point* of A. If A is identical with its set of limit points,* it is said to be *perfect*. Finally, if for any two points of A and any distance d, there exists a sequence of points of A linking these two points to each other by a sequence of distances each of which is less than d, we shall say that the set A is *cohesive* [*d'un seul tenant*].

A cohesive perfect set is a continuum: volume, surface, or line. But surfaces and lines do not have interiors; each of their points is a boundary point, that is to say, a limit point of points which are not members of it. On the contrary, the boundary of a volume circumscribes the points of the volume which do not happen to be boundary points. We shall therefore say that a *cohesive perfect set of points* A *is a volume if any point of* A *which is a limit point of points outside* A *is a limit point of the points of* A *which are not themselves limit points of points outside* A.[(2)]†

Such is volume in the geometry of points: a class of points connected by certain distance relations. We can now construct relations between volumes from relations between their points: for example, we can agree to say that one volume contains another if every point of the latter is also a point of the former, and that two pairs of volumes A, A′; B, B′ are *conjugate* if there exists a point of A and a point of A′, a point of B and a point of B′ which are separated by the same distance [(3)]; this is the relation which corresponds to the experi-

(1) The interior of a sphere is easily defined: it is the class of points which are closer than the points of the surface of the sphere to its centre. The interior of any particular volume is also easy to define – the difficulty lies in framing a *general* definition of volume.

(2) If we wish to avoid having to regard two volumes connected by a point as a single volume, we must add the condition that any two points a, b of a volume belong to a coherent perfect set all of whose points apart from a and b are interior to it.

(3) We can also say that three volumes are *aligned* if some one line crosses them – in the world of experience, it is a sighting linking three landmarks which corresponds to this relation.

* Translator's Note: As it stands, this condition defines what in modern terminology is called a *closed* set. A set is then said to be *perfect* if it is closed and every point of it is a limit point of *other* members of it. I assume this is what Nicod intended.

† Translator's Note: *i.e.* if, in modern topological terminology, the boundary of A is a subset of the closure of the interior of A.

mental measurement of the distance between two bodies by placing one end of a ruler inside one of them and the other end inside the other.

Let us now wherever possible insert the expressions just defined into the theorems of the geometry of points: certain statements no longer explicitly contain points or relations between points, but only volumes and relations between volumes. Let us make a separate list of these statements in their new form and then take them on their own merits. In other words, let us stop describing volumes in terms of points, and instead let us take *volume* and the relations of inclusion[1] and conjugation of volumes as primitive expressions with entirely undefined meanings. Let us call this set of propositions, which can be divided into axioms and theorems, the *properties of volumes.* These are properties which seem likely to adapt themselves more directly to nature.

Do the properties of volumes comprise the whole of geometry? Do these properties, which are formally included in the properties of points, formally include the latter? In the point system, it is possible to construct sets of points which obey the laws of volumes – and this is how we obtained them. But is it possible, in the volume system, to construct sets of volumes which obey the laws of points?

We present Mr Whitehead's solution.

Turning again to the geometry of points, we show that there is a unique class of volumes associated with each point, and that to the fundamental relation of congruence of pairs of points there corresponds a unique relation of the pairs of sets of associated volumes. Each axiom of congruence of points is then translated by a property of this relation in the geometry of volumes: the sets and relation considered thus form a 'solution' of the geometry of points.

The set of volumes associated with each point is simply the set of all the volumes containing that point. But we may not define it in this way, for its definition must mention volumes and relations between volumes *only*. We must adopt a very roundabout method.

If we wish, by means of relations between volumes, to define a

(1) It would have been more correct to adopt, in place of *inclusion*, which has a purely logical meaning, an entirely isolated term such as *absorption*, for example. However, we have found it preferable to retain the standard word, with the understanding that it here denotes a variable, and, as a result, quite indeterminate, relation.

class of volumes which have only a single point in common, we immediately imagine [a class of] volumes containing one another and becoming smaller than any given volume, that is to say, a class of volumes such that, given any two members, there is one which contains the other* or is contained by it, and further that there is no volume contained in all its members: let us call such a class an *abstractive class.*

But these conditions are not sufficient to ensure that all the volumes of an abstractive class have only one point in common: they might have not a volume but a line or surface as their common nucleus. Thus, an abstractive class composed of thinner and thinner discs of the same diameter has for its nucleus a circle; an abstractive class formed from thinner and thinner cylinders of the same length has for its nucleus a segment of a straight line. Let us, therefore, attempt to obtain the additional condition which will ensure that [the members of] an abstractive class have exactly one point in common.

First of all, we must impose a restriction which does not make its appearance in Mr Russell's[1] exposition, but which nevertheless is essential for the accuracy of the solution which follows: we must confine our attention to those abstractive classes whose common points are *interior to* (and not on the surface of) all the volumes of the class. This preliminary condition is satisfied if we stipulate that, of any two volumes of the class, one must always contain the other *without tangency.* But what is this new relation between two volumes?

When volume A, contained in volume B, touches the surface of B not at a point or in a line, but in a portion of a surface, a volume C of which part is exterior to A can penetrate A and yet not contain any volume contained in B but not in A (Fig. 1): this is the acting definition of the 'inclusion with surface tangency' relation between volumes.

We now proceed to the general case in which the tangency can be linear or at a point. The volume A, contained in B with contact, forms the common nucleus of a nest* of volumes A′, A″, each of

(1) It also cannot be found in the more recent works in which Mr Whitehead extends the method of abstractive classes to the definition of an event-point of space-time, in spite of the fact that it seems necessary there too.

* Translator's Note: In modern terminology, a class satisfying this condition is called a *nest, tower* or *chain.*

which is contained in B with surface tangency (Fig. 2): this then becomes our general definition of inclusion with contact, and hence of inclusion without contact. Let us term *interior* abstractive classes those classes of volumes which are contained in one another without tangency, and let us restrict ourselves to interior abstractive classes from now on.

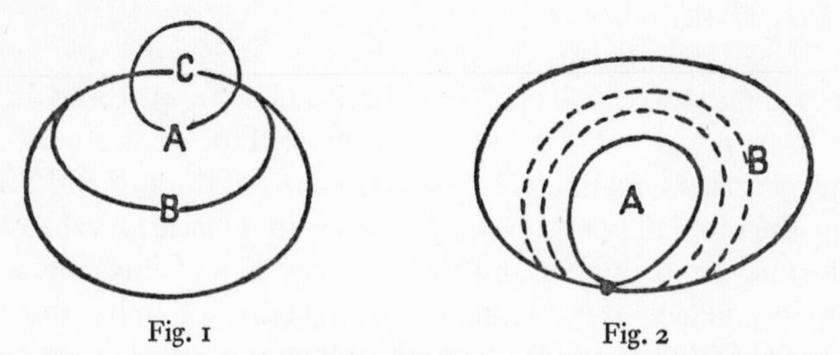

Fig. 1 Fig. 2

Consider one of these, say A, which has as its nucleus a piece of a surface or of a line, and another, say B, whose nucleus, which can be either a smaller piece or a single point, is part of the nucleus of the first.

Each volume of A then contains a volume of B; we express this state of affairs by saying that the class A *covers* the class B. On the other hand, as soon as the volumes of B crowd close enough together around their common nucleus, leaving outside of themselves part of the larger nucleus of A, they therefore no longer contain any of the volumes of A: therefore the class B does not cover the class A. Hence each interior abstractive class whose common part contains more than one point covers every interior abstractive class which does not cover it.

Let us now consider a class A whose common part contains exactly one point p, and suppose that B is a second class covered by A; the common part of B, which is contained in that of A, also reduces to the single point p. Consequently, each volume of B contains a volume of A which grips the point p harder than it does and so conversely the class B covers the class A.

Each interior abstractive class whose common part reduces to a single point is therefore *covered by every interior abstractive class it covers:* here is the condition we were looking for. Let us call the abstractive interior classes which satisfy this condition *punctual classes.*

Each punctual class of volumes is associated with the unique point which forms its common part. But the converse does not hold, because each point is the nucleus of a punctual class of spheres, or cubes or cylinders, and so on *ad infinitum*, and each of these punctual classes covers every other. However, we can unite them into the collection of all volumes which are members of punctual classes covering a certain punctual class. (This collection includes the latter in view of the fact that every class covers itself.)

This collection, which Mr Whitehead calls a *punctual element*, comprises the unique class of volumes to be associated with each point in the geometry of points; it is indeed the class of volumes which contain this point in their interiors, but it is defined only by the relation of inclusion between volumes.

This is Mr Whitehead's construction, or, to be more precise, this is what it becomes in passing from the analysis of the real world to the realm of pure geometry, in which place it appears naturally situated.

It still remains to establish a correspondence linking the fundamental relation of congruence between pairs of punctual elements associated with them. This relation is obtained by conjoining all the pairs of volumes which are respectively members of the two pairs of punctual elements under consideration; let us call it *equivalence* of these two pairs of elements.

Now we have used the phrase 'conjugation of the pairs of volumes X, Y; X′, Y′' to express the fact that there exists in these volumes two congruent pairs of points x, y; x′, y′: clearly, if the punctual elements A, B, A′, B′, have as their common part the points a, b, a′, b′, respectively, and if these latter form two congruent pairs when taken in this order, then any couple of pairs of volumes X, Y; X′, Y′, members of A, B, A′, B′ respectively, will be conjugate. Conversely, if every couple of pairs of volumes X, Y, X′, Y′ taken from the punctual elements A, B, A′, B′ respectively, contain two congruent pairs of points x, y; x′, y′; then the nuclei a, b; a′, b′ of these elements form two congruent pairs themselves.

Each axiom of congruence of points is thus at a much greater distance reflected in the geometry of volumes by a property of equivalence of punctual elements. Let us now isolate this geometry: let us regard the *volumes* as no more than a class of undefined terms, and their relations of *inclusion* and *conjugation* as being mere relations of these terms, also undefined.

Let us compare this system with its forerunner. We reaffirm that each of these systems can be interpreted in the other if we write, in the point system,

volume = coherent perfect set of points each of whose boundary points is a limit point of its non-boundary points,

inclusion of the volume A *by the volume* B = identity of each point of A with some point of B,

conjugation = relation between pairs of volumes containing respectively two congruent pairs of points,

and, conversely, in the volume system,

point = punctual element = interior abstractive class covered by all interior abstractive classes which it covers.

congruence = relation between two pairs of points A, A′; B, B′ each of whose volumes a′, a; b, b′ form conjugate pairs.

The geometrics of point and volume are thus *inseparable*. Each satisfactory interpretation of point and of the relations between points furnishes a satisfactory and more complex interpretation of volume and the relations between volumes, and each satisfactory interpretation of volume and of the relations between volumes furnishes a satisfactory and more complex interpretation of point and the relations between points. Hence, geometry does not in any way demand of nature volumes made of points rather than points made of volumes; it adapts itself equally well to simple points or simple volumes, to a world in which there are only points or a world in which there are only volumes.

Let us imagine such a world, in which volumes and their fundamental relations alone have simple meanings. The geometry of volumes is the only one appropriate. But the geometry of points would appear to the scientists of this world as a refinement of the geometry of volumes; it would be a construction of extreme elegance. Through the introduction of the involved concept of punctual element, it would wonderfully simplify the statement of the laws of nature. And if we suppose that this world does not lack metaphysicians, they are bound to dream that the punctual element is the manifestation of a simple entity, real or ideal, and thus gratuitously to infer the simplicity of terms from the simplicity of laws.

But let us, by means of an example, demonstrate the simplicity communicated to the laws of volumes by the concepts of abstractive class and punctual element. Consider the relation between two conjugate pairs of volumes. This relation does not enjoy the simple property of transitivity, but it does possess the following more complex property: if the two pairs of volumes a, a′; b, b′ are both conjugate to the pair of volumes c, c′, one of them, say a, a′, is always conjugate, not to the pair b, b′, but to a certain pair of volumes x, x′; which intersect b and b′ respectively and are smaller than c and c′ respectively (the intersection of two volumes being the inclusion in both of some one volume, and the volume m being smaller than the volume n if n contains a pair of volumes which are not conjugate to any pair of volumes contained in m). We can express this in an imprecise, but familiar, way by saying that two pairs of volumes conjugate to the same pair are conjugate to each other with a possible error equal to the size of the volumes of the intermediate pair; an imprecise statement, in fact, in view of the fact that the law does not say that these two pairs are conjugates, and there is no mention of error.

But let us take, instead of the conjugation relation between pairs of volumes, the relation of equivalence between pairs of punctual abstractive classes formed by the conjugation of all the pairs of their members: the complicated property possessed by the simple relation of conjugation is seen to communicate the elementary property of transitivity to the complex relation of equivalence, as is easily shown.

Is there a better example of simplication of laws through complication of concepts? In this respect the point plays in the geometry of volumes a role similar to that of energy in mechanics, or entropy in thermodynamics – complex entities whose simple laws elegantly *express* the science.

There is a notion which can only be approached in anticipation, since it only assumes its proper place at an advanced stage of research we shall not attain. But it springs to the mind spontaneously and troubles it. Let us examine it for a moment.

Does nature, which does not supply points, really supply volumes?

It appears to provide only a lame interpretation of these latter for its 'volumes' are not sensible unless they are sufficiently large. To assert that they satisfy geometry is again to proceed beyond the data. Hence do we gain anything by abandoning points in favour of volumes? Does not the application of geometry to nature, in whatever form it may be expressed, always require the same rash assumption?

No: because the geometry of volumes, as opposed to that of points, has at least one part which is illustrated in nature, and depends for the rest on an assumption which is much preferable.

Indeed, if we admit that there is a natural interpretation for volumes and their relations above a fixed magnitude, all the propositions which do not entail the existence of volumes smaller than a given size, will appear as laws of nature. Thus the theorem just stated which was concerned with the 'approximate' transitivity of the conjugation relation between pairs of volumes would be a rule precisely verified in experience, for it is important to observe that the truth of a property said to be approximate is something quite different from the approximate truth of a property said to be exact. The second is merely a confused vision of the first, which alone is something rigorous and satisfactory. Such is the relation, in the world of experience, between the geometry of points and that part of the geometry of volumes which does not postulate the existence of arbitrarily small volumes as small as one likes, which might be called its positive part.

But even the hypothesis asserting the presence of such volumes, which introduces the punctual element, differs radically in character from the hypothesis positing the presence of simple indivisible points. Indeed, instead of postulating the existence of entities of which there are no examples in nature, we confine ourselves to positing new members of a known class, which differ from the known members only in the way these differ amongst each other: an intelligible and modest hypothesis. It arises naturally from the positive part of the geometry of volumes: for we have seen that it communicates a simplicity to the laws of this geometry which cannot fail to strike us.

But again, is not admitting into nature sequences of nested (*emboites*) volumes, which decrease without limit, the same as positing simple terms towards which these sequences converge? One of the merits of the study of geometry in its empty form, independent of any meaning assigned to its expressions, is that it exposes the unsound nature of this inference. For it is clearly worthless in the general case: since we do not know what a volume or the inclusion of a volume by another may be, what grounds would we have for asserting that any punctual class, i.e. any class of volumes which contain one another without tangency, which do not contain a volume in common, and which furthermore is such that any one which is contained in some volume of each similar class any volume of which is found to be contained in some volume of the first, I repeat, what grounds would we have for the assertion that the class of volumes just defined necessarily provides, like a bolt from the blue, a new term, of a quite different nature from the volumes composing it? There is consequently no logical necessity here. If sequences of volumes of this type cannot exist in nature without simple terms of convergence, it would only be by virtue of a contingent property of natural volumes, a property which is doubtful anyway and whose absence would not affect at all the application of geometry to the world.

*

We have attempted to grasp geometry in its abstract sense. We have assumed the wholly formal point of view of a man who, not knowing the meaning of the characteristic terms of this science, is ignorant of what objects or relations it claims to discuss, and has not the least idea about them. The reading of a geometry book is then nothing more than an exercise in logic, which consists in verifying that the theorems are genuinely derived from the given axioms.

Is it therefore possible to follow a proof without knowledge of the subject matter it discusses? Indubitably, for a proof is completely independent of the meaning of its terms. To be able to recognise that all Frenchmen are mortals, if all men are mortals and all Frenchmen are men, there is no need to know what a Frenchman is, or a man or a mortal. It is equally unnecessary to know what a point, or a line, or congruence, is, to be able to grasp the force of a correct demonstration, or even to appreciate its elegance. There is no geometer of our time who will not grant this.

But now that we have turned the last page of this book on an unknown science called geometry, which chance has placed in our hands, when we have checked the links in its proofs, and sufficiently admired the necessity and ingenuity of the reasoning which, by the development, *ad infinitum*, of the theme proposed by the axioms, bind together in a thousand ways a few expressions which are devoid of meaning to us, our curiosity takes a new direction. We ask ourselves whether there really are entities to which these 'geometric' expressions apply, or rather *could* apply with accuracy. The set of axioms which begins our book thus becomes the datum for an entirely new problem, in which the geometric expressions form the unknowns: *Are there meanings which, if ascribed to the geometric expressions appearing in these axioms, transform them into truths?* Each system of meanings which resolves this problem is an illustration, an example, an interpretation, or, better still, a *solution* of the set of axioms concerned.

About these meanings the axioms themselves say nothing. They allow us to ignore their nature and the degree of their simplicity or complexity. Moreover, they perhaps admit diverse independent solutions of extremely contrasted types. It is, therefore, advisable to refrain from postulating the existence of a pre-eminent geometric domain in each determinate region of the real or ideal world.

But geometry can be indiscriminately based on the most dissimilar groups of axioms, without a common primitive expression appearing in all of them, not even 'point'. We have defined the purely formal equivalence which transforms these systems into the facets of a single geometry; and we have specified the sense in which their solutions, although different, are logically inseparable. Let us agree to call any system of meanings satisfying the axioms of a geometry a *space*.

We ask ourselves whether spaces exist. They do; we have come across one in the domain of numbers. Hence we know that geometry does not conceal any contradictions, since it is not without examples. These arithmetic solutions, being the only ones which exist *a priori*, are the only theoretically certain solutions. However they do not interest us, for it is not in the domain of pure ideas, but in sensible nature, that we wish to see geometric order reflected. With this order in mind, let us turn to sensible nature, so as to find out whether it too does not offer one or several illustrations.

Part Two

SENSIBLE TERMS AND RELATIONS

Introduction

We must now take cognisance of the elementary objects – terms or relations – which sensible nature offers us. These terms and relations form the fibre of its structure; it is in the textures they create that the features of geometry should reappear. But to speak of nature as a fibre of terms and relations is enough to arouse all the suspicions of the philosopher. Is not this a considerable postulate? Who gives us the right to apply the categories of logic to the sensible process? There are those who feel that reason misleads us here.

So deep-rooted a suspicion cannot be attacked head on. But let us proceed: as we enter into detail, we shall see the grounds for this suspicion assume precise outlines, and then, perhaps, vanish.

The elementary terms of nature are entities known as *sense data*. These are what we refer to as *this* when we say to ourselves, in speaking of something immediately present to one of our senses, *this is a tree*, *this is a penny*(1), or again, *this is a shooting-star*, *this is the song of a nightingale*. For logic, which is ignorant of time, knows neither flux nor rest, and a sense datum need not be free of change for it to be a 'term'. But how can the sensible process in which everything is fused be resolved into distinct and simple terms? We shall perhaps see this more clearly in a moment. Let us proceed to examine the elementary relations our mind can distinguish among these terms whose nature will become clearer as we progress.

(1) Cf. G. E. Moore, *Aristotelian Soc.*, suppl. Vol. II (1919), page 180.

CHAPTER I

Spatio-temporal Relations Independent of the Distinction between Extension and Duration: The Notion of Sensible Term

The relation of spatio-temporal interiority. I follow with my eye the flight of an eagle crossing my field of vision in a slow and continuous glide, the whole of which I perceive as a single visual term. In the middle of its flight, the eagle flaps its wings once. Between the one event, namely the flap of the bird's wings, and the other, larger, event, namely its flight, I perceive a very clear and doubtless very simple relation which I express by saying that the first of these two sense terms is *interior* to the second. With my eyes closed, I slide a pencil over the fingers of my open left hand. Between the one event, namely the transit of the pencil over my index finger, and the other, larger, event, that is the transit of the pencil over my whole hand, I once again perceive a very distinct relation which seems to be the same as the previous one, and which again leads me to say that the former term is *interior* to the latter. It appears to be the same relation as that which I apprehend between the sound of a word and the sound of the sentence in which it occurs, between a patch of a painted figure and the larger expanse formed by the complete picture in which it appears. This relation is distinct and obvious and springs to the eye in each case.

Its connection with spatial and temporal inclusion. Although it does not as yet distinguish between *temporal* and *spatial relations*, this fundamental relation of *interiority* involves simultaneously both the durations and the extensions of the sense data which it connects. For every sense datum which is *interior* to another is evidently *included* in that other both from the *durational and the extensional points of view*. But we must define, or rather indicate, these two more special relations. The relation between two of my sense terms *a* and *b* which holds when the duration of *a* is enclosed within the duration of *b* I call *temporal inclusion*. So, during an August night, I can see one shooting-star materialize and then die during the somewhat longer life of another in a different part of the sky. The first flash of light is then temporally (but not spatially) included in the second. In contrast the relation between two of my sense terms *a* and *b* which holds when the immediate extension of *a* is enclosed within the immediate extension of *b* I call *spatial inclusion*. Thus, from the standpoint of a motionless person watching a fire, the small grey patch formed by the ash at the end of the fire is spatially (but not temporally) included in the large patch made by the flames a little while before.

From what I have just said, it might appear that the relations of spatial and temporal inclusion only connected my sense terms through the agency of other terms, namely their *durations* and *extensions*. We shall return to this problem later. However, let us say at once that when speaking of *durations* and *extensions* of sense terms belonging to the *field of sensation*, and more especially to the *field of a particular sense* and to the different *regions* of this field, I shall merely be simplifying the language, and really mean to discuss no terms outside my sense data, and no relations apart from those I perceive to hold among these data.

Interiority entails both *spatial* and *temporal inclusion*: any sense term interior to another has its extension and its duration respectively enclosed within the extension and the duration of the other.

Would the converse hold too? For one term to be interior to another, is it not sufficient that the first term occur within both the extension and the duration of the second? Spatial and temporal inclusion would then form a necessary and sufficient condition for interiority. Hence, the relation of interiority between sense terms might not be the simple relation we imagined it to be. If it is really equivalent to the conjunction of spatial and temporal inclusion, interiority must be identical with this conjunction.

Of course, this analysis could always be rejected, for two notions can be inseparable with respect to their instances and yet remain different: for example, 'equilateral triangle' and 'equiangular triangle'. In the last resort, it is always by direct inspection that we pass judgement on the reduction of one notion to others. Now it seems to me that, in the relation between the flap of the wings and the flight, between the figure and the picture, between the word and the sentence, I find a simple connection which does not entail the similarity and the conjunction of a relation in respect of extension and a relation in respect of duration.

But is it even true to say that interiority is inseparable from this conjunction? Indeed no: when moving data are involved, temporal and spatial inclusion together no longer imply interiority.

Fixed and moving data. The distinction between *fixed* and *moving* sense data is a directly perceived qualitative contrast. We therefore cannot define it; we can do no more than indicate it by means of expressions which contain it. A datum is fixed if throughout its duration it maintains a constant extension, and remains in the same position in the visual field. Conversely, a datum is moving if during the course of its duration its immediate extension varies either by deformation or displacement.

Reason obviously cannot approach sensible motion with the purpose of making it a decomposable entity. For logic knows not change or constancy, motion or rest, just as it is unaware of colours and favours none of them. A patch is moving or fixed just as it is green or blue, and to the intellect, sensible motion may be as simple a quality as rest.

Now interiority applies to both fixed and moving data, and is quite indifferent to their distinguishing features. Whether the sensible event is completely quiescent or full of movement, whether it is a landscape or a battle, interiority always relates a detail to the whole in the same way. The flap of the wings is interior to the flight, just as the figure is interior to the picture, and the word interior to the sentence, each one of the names here denoting, not the actual thing, but the sense datum presented on a certain unique occasion. Temporal inclusion also does not distinguish between fixed and moving terms: they all endure in the same way; they can include and be included in one another with respect to their duration. Spatial inclusion on the other hand raises problems where moving data are concerned: indeed, what does it mean to be enclosed within, or to enclose, the extension of a moving datum, since a moving datum is precisely one of variable extension? Now either the question is unanswerable, in which case we shall conclude that so far as moving terms are concerned, interiority, always a clear and manifest relation, cannot be reduced to the conjunction of two relations one of which now has no meaning. Or else the question can be answered: we shall see, however, that the conclusion will be the same.

Essentially, what is the relation of spatial inclusion? It is one sense datum saying to another in my mind: 'You have been nowhere I have not been in my time; your whole domain has been mine; you have never escaped from the shadow I call my extension.' But from this viewpoint, the extension of a moving term consists of the *whole region of the field of sensation 'swept out' by this term in the course of its existence*. For this is the domain on which it has impressed its quality, this is the path it has traced out, and over which its memory holds sway. Thus the extension of a shooting star is the complete line it traces in my visual field.

Spatial inclusion therefore would also apply to moving, as well as to fixed, terms. *But its conjunction with temporal inclusion no longer entails interiority*. For instance suppose a moving sense datum, for example a cloud which I see crossing the sky framed in my window. Its extension consists of the whole band it has swept out in passing. But it does not pervade its extension as does a fixed datum; it is possible that a second, briefer and more restricted datum – for example a star's twinkle – may arise within the extension and the duration of the cloud, and yet be exterior to it.

Thus both the facts and intuition apparently lead me to recognise that the interiority of sense terms is a simple relation which, while entailing spatial and temporal inclusion, is yet not entailed by them, and that it is a more concrete, more undifferentiated nexus, which is antecedent to the division of relations with respect to extension and duration. The relation of interiority is of such import for the concept of sense-term that it is best to stop for a moment to discuss its nature.

Interiority and logical composition. Does this relation not possess a truly rational meaning? Does it not involve the *logical relation* between *component* and *composite*? This question is of great importance: for if the interiority of one sense term to another implies the logical relation between part and whole, then the complexity or simplicity of a sense term depends on the number of terms interior to it. As a result, the smallest terms enjoy the privilege of being real.

In the eyes of reason, experienced reality is, so to speak, precipitated as a fine dust composed of sensible points and instants, for composites only achieve reality through their elements. Moreover, this view is in accord with the fundamental scientific principle which claims that physical reality is entirely determined by its state at each point and each instant. However, we must investigate this question more deeply.

The criterion for logical simplicity of a term. We must first investigate more precisely what a simple term means, intellectually. We can establish the following criterion. Everybody will agree that when a term or content *x* is *part* of the term or content X, it is *impossible to conceive something about* X *without conceiving the same thing about x*. On the other hand, if I can make an assertion about X without thereby asserting anything about any other term whatsoever, then it is clear that X is intellectually a simple subject.

This is not a mere matter of words. For the mind at times recognizes simple terms which have no names in language: indeed, every simple content uses *this* as its first name. Conversely, a single word sometimes unites a mass of details which form only a mass and not a whole. The verbal expression of judgements does not indicate with certainty to which simple subjects they apply, for many words can cover either a simple content or a complex one, according to the permanent or ephemeral attitude assumed by the mind which employs them. Thus it is conceivable that a common part of our lives to which we assign the same name, for example, a walk taken together, may have been for one of us a succession of events, and for another, a single event. Perhaps even the person who recollects it as a succession of events may retrospectively discover an aspect of sensible unity in a moment of particularly clear perception; and it is possible that the ones who had at first apprehended the total aspect may sometimes lose this faculty through some kind of fatigue of the imagination. What was at first a single *this* may become at certain moments no more than *this* and *this* and *that*; fluctuations of the objects apprehended by the mind is not conveyed by words.

The logically simple or complex character of these objects is no longer determined by the psychological conditions for their presence. Indeed, let us suppose that an object possessing property

X is mentally inseparable from objects possessing properties x_1, x_2, x_3, . . ., and that these latter properties are connected by the relations R_{12}, R_{13}, R_{23}, ... The desire for economy inherent in reason inclines us to say that the first object is nothing more than the others 'taken together', and that the property X is resolved into the complex property derived from the properties x_1, x_2, x_3, . . . and their mutual relations R_{12}, R_{13}, R_{23}, ... But the principle of economy, under the pretext of stopping us from seeing double, must not blind us in the process. It is direct inspection which in the last resort decides whether property X is *through its very meaning* a distribution of properties and relations to a collection of subjects, or if, on the contrary, by virtue of its simplicity, it requires a simple subject.

Application of this criterion to large data. Do we necessarily find this logical relation of composite to component between a sense term and those terms interior to it in duration and extension, or between the patch formed by a chess-board and the patch formed by one of its squares, or between a chime and one of its individual notes? Obviously not! For there certainly are some judgements about the chess-board, or the chime, which predicate nothing of the squares, or of the notes. If I judge that the chess-board has a square appearance, I ascribe a certain property to the entire patch presented me, and not to the smaller patches interior to it, although each of them, taken on its own merit, presents the same property precisely because of this. Indeed, if my judgement applies to the square as it does to the whole chess-board, it follows that in regard to the chess-board it does not distinguish the squares on it.

If we pass from the spatial aspect of interiority to its temporal aspect, the situation is once again the same. The chime, like the chess-board, can be the simple subject of a judgement. Such a clear illustration is not possible here for we cannot distinguish a property common to both the total sound of the chime and the sound of one of its notes. Nevertheless, we shall admit the possible individuality of the complete sound, because its melody may appear to be a combined property flowing over the partial sounds without attaching itself to them. Here the possession of an unanalyzable property has been posited as the criterion for logical individuality.

The relation of interiority does not imply the logical relation of component to composite, and the most extended and prolonged sense data of the richest internal variety may appear as simple terms in the light of reason.

But why is one deceived? Why does interiority, a purely empirical and contingent relation between two terms, apparently effect an analysis of one term for the other's benefit?

Actual determination of large data by restricted data. Sense data are not composed of the sense data they may enclose. They are distinct entities with simple characters of their own. However, let us investigate what *determines* these characters. Does not the square appearance of the chess-board when viewed from the front as well as the presence of melody in the sound of the chime, admit a proximate cause? Certainly: the presence of these qualities in the complete expanse, or in the complete sound, is determined and guaranteed by the presence of certain qualities and relations in the smaller patches, or the shorter sounds, which are interior to them. Thus the chess-board looks square when viewed from the front *because* (but not *in that*) this object comprises precisely eight columns and eight rows, each of which contains eight square compartments. Indeed, for a coloured expanse to appear square, it is sufficient that it satisfy these conditions.

In the same way, the melody I recognized in the total sound of the chimes is present in every sound which contains in it other sounds possessing certain definite qualities and relations symbolized in musical notation. It is doubtless permitted to generalize these two examples and lay down the principle that *all properties of sense data which contain others are determined by the properties of the latter.*

Give me a detailed account of the notes and their relations, and you give me the melody; give me the colour of each individual point of the canvas, and you give me the whole picture. One now perceives that sense data relatively large in extension and duration, such as the sound of a melody, or the variegated scene of a picture, come to be regarded as simple aggregations of more restricted terms. But then one tells oneself: If the larger terms are determined by the smallest, it must be the case that these former have no independent reality. However, this is a confusion. The properties of restricted terms *determine* those of large terms, but do not actually *constitute* them. Knowing the former, I can infer the latter only if I am already acquainted with them. Let us imagine a mind incapable of apprehending larger sense-data in their entirety: it will never apprehend the larger terms, the whole texture of which is constituted by those which it perceives, the pictures of which it sees the brush-strokes but does not see, the songs of which it hears the notes but does not hear. Let us now endow it with a more synthetic outlook: by embracing larger portions of the 'same' sensible process, it will discover new and more comprehensive, but not less simple, entities whose original qualities, although determined by the qualities and relations of the more restricted terms they contain, nevertheless reside nowhere but in themselves.

Moreover, is it reason which affirms the principle that the properties of a sense term are dependent on the properties of the terms interior to it? No, it is merely an empirical law. Of course, we would be more surprised to find it falsified than we would be to hear a tree speak. But no matter how strongly habit inclines us to accept it, reason still remains indifferent. It perceives nothing here but a fact, a relation between *existences* and not between *essences*. In its eyes, the restricted or large data which overlap in the sensible process in so many different ways are all real and simple in the same sense.

The philosophical defence of large data. We have thus removed the conflict which has sometimes placed the philosopher, who defends the realities of life and art, in opposition with the logician and physicist. In fact, it is data of relatively large extension and duration which give experience its rhythm and organize it, so to speak. Common sense must regard with suspicion the disintegration through the preceding confusion of these natural, almost living, terms into a mass of sensible minima. But it does not see where the error lies, and is itself deceived, for how can we be expected to believe that what we see of a house is not composed of what we see of its stones? The philosopher sometimes entertains this obscure protestation and attempts to meet it. The sense terms we find important, he says, are individuals in spite of their broadness and diversity. Moreover, if the ordinary man grasps a chess-board as a single visual term, does not the painter apprehend a vast landscape in the same way? Does the musician not perceive the whole composition to be a single resonant term? It is only the first step that is difficult: do not days, and even years, have their own physiognomies which are unveiled at certain privileged moments? Can I not, in view of this, conceive of my whole sensible past as forming one single event, and my whole experience as forming one single term which expands continuously and indivisibly?

There is a contrast between the technician's analytic attention, directed towards sensible details that are difficult to grasp owing to their minuteness, and the artist's synthetic attention directed, on the contrary, to broad and rich terms whose apprehension is made difficult by the richness and breadth of their extension or duration. At one extreme we have the discernment of point-instants, and at the other the apprehension of all experience as one single term.

Here occurs the confusion alluded to a while ago. Indeed, if we admit that a sense term cannot be interior to another in extension and duration without in the eyes of reason forming one of its parts and that, consequently, the former must possess a more fundamental reality than the latter, then we must decide once and for all between restricted and large terms; we must ascribe experienced reality to one or the other of them, either to the terms of the technician, or to those of the artist, and no longer to both at the same time.

Once involved in this dilemma, we may either decide in favour of analysis as did Leibniz, in which case the reality of sense terms will crumble into dust; or, as Bergson does, in favour of synthesis, so that reality will only belong to the totality of immediate experience. Or else, as does M. Bergson on other occasions, sensing the mind's discomfort when confronted with both extremes, we may accuse reason itself for foisting a deceptive choice on us.

But we only mislead ourselves by making reason intervene in situations to which it is indifferent: this mystery of the sensible whole not being the sum of its parts vanishes as soon as we realize that these are not true parts, and that interiority in extension and duration does not constitute a rational relation.

Other relations in the same family. Other relations in the same family as interiority are *interpenetration*, *exteriority*, and its limiting case, *continuity*. A row and a column of a chess-board are examples of two sense terms which *interpenetrate*. They have a common interior term, but my apprehension of their interpenetration seems to be something simpler and more primitive than discernment of the square in which they intersect. Likewise the sound of the three first and three last lines of a four-lined stanza form two interpenetrating sense terms. I say, on the contrary, that two lines of one strophe, two rows in a chess-board, two actions separated by a pause, are two terms *exterior* to one another in as much as they do not interpenetrate. Finally, two terms exterior to one another in this sense are said to be *continuous* if they touch each other without penetrating, like the motions of two relay runners, one of whom starts as soon as the other arrives.

To all these relations we can apply the same arguments which led us to recognize interiority as a simple relation which is irresolvable into relations of extension and duration, and also as a purely existent relation with no rational import. Just as a sense datum can be a logically *simple* term while containing many others, so a dense datum can be a logically *distinct* term although it interpenetrates many others and draws on them indiscriminately. Reason no more circumscribes the sense data it distinguishes than it breaks them asunder; it accepts their lack of clear boundaries in extension and duration.

The indefiniteness of sense data. In what does the indefiniteness of a sense term such as a cloud, consist? It consists in the fact that there are other clouds which I cannot definitely classify as exterior or non-exterior to the first. But this uncertainty does not at all destroy the individuality of the first cloud as a sense term. For how can we say that the apprehension of any two individual terms enables us to discern with certainty whether they have such and such a particular connection or not? We do not infer from our inability to say whether two given coloured patches have the same shade or not that these two patches are not distinct individual data. What gives us the right to draw the opposite conclusion from an entirely similar uncertainty concerning exteriority in extension and duration? It is much more worthwhile to notice that in this respect uncertainty has its limits. There is no sense term so blurred that we cannot discern in respect of some other terms, whether they are exterior to it or not.

No matter how indistinct may be the visual datum which I distinguish by remarking 'What a beautiful cloud', I am none the less sure that this pebble is exterior to it. However diffused the sunset there comes a time when night has certainly fallen. From the standpoint of the relations of interiority and exteriority, there are, then, for every sense term, other terms which can be precisely classified with regard to the first. Now, while the existence of doubtful classifications does not preclude the distinct and individual mental presence of the term considered, the existence of certain classifications would seem to establish this presence.

Absolute reality of sense terms and relations. There are no hidden flaws in the procedure I use to discern, among the flux of my sensations, the terms connected by relations of interiority, exteriority or interpenetration. Nothing compels me to ascribe to these relations a rational meaning they do not in fact possess, so condemning me to contradiction. When I say to myself *this*, when I distinguish such and such a datum in order to say such and such a thing about it, I do not abstract it from the continuous stuff of my experience, I do not arrest its development, I do not raise it above the flow of which it remains a fleeting wave. By regarding it as a logical term, I do not assume that it has a more real unity than the restricted data it contains, nor the large data which contain it, nor those which bite into it and otherwise divide experience. There is in the flux of my sensations a multitude of realities, a surprising wealth of entities that interpenetrate without losing their original simple quality. But this is not a kind of miraculous vision which unveils itself for intuition alone and vanishes when confronted with reason. Of course, this interpenetration of simple realities would be a miracle if it were necessary to take it in a purely logical sense; but in fact this is not so. The contrary view rests on an inadequate grasp of logical abstraction.

The contrary view may, however, be retained. But let us now assume this contrary point of view. Suppose we admit the existence of a hidden flaw in the discernment of distinct and related sense terms. At first all seems chaotic: but the situation is like a painting of a shipwreck: for here we are reinstating what was apparently to be rejected. Without a doubt, we are told, there are in a *certain sense* distinct and related terms in the flux of sensation since the physicist, the chemist, and the astronomer observe them. Moreover their pronouncements are not vitiated by error in the ordinary sense, but only by that metaphysical error which is related to intellectual weakness rather than falsehood. They do not live in the realm of the false, but rather in that of the artificial, the superficial, the symbolic – terms which lay claim to a certain glamour. These words, which waver between error and sin, between truth and the intrinsic merit of thought[1], form the current coin of metaphysics. They enable it to destroy and preserve everything. These are tricks of the trade, and this case is no exception. It is, therefore, sufficient to answer: 'Since, according to what you say, it is not true that distinct and related sense data exist, and since it is not true that lightning precedes thunder and yet true in a certain sense, you must take what we say about them in this sense, whatever *that* may be, for the remarks we make about them will always be of this nature.'

But, for the scientist, is such detachment justified in philosophical investigation? Yes, because the independence of two problems is a fact – not only with regard to our ignorance, but in itself. Of course, the philosopher refrains with difficulty from saying with Descartes: 'All my opinions hang together; you must accept them all or reject them all.' For are not our most cherished beliefs always the ones least likely to be true? However, this excessive cohesion is most often illusory. Today the idea is spreading that philosophy will progress only by becoming more fragmentary after the fashion of the sciences. The philosopher must learn to divorce himself as far as possible from his own doctrines when engaged in a particular problem. He must present its solution in a language, and if possible in a spirit, as independent as possible of all his other beliefs.

(1) Cf. H. Bergson, *Données immédiates*: 'A definition of that sort contains *a vicious circle, or at least a very superficial idea of duration*' (page 76).

CHAPTER II

Temporal Relations and the Hypothesis of Durations

I now propose to pass from interiority, interpenetration, exteriority, and continuity, which clearly form a first group, to a second family, namely that of the *temporal relations* among my sense terms.

We have already encountered temporal inclusion: it is nothing more than the relation *during*. Let us consider in addition the following: *interference*, the relation between two data which holds when one begins before the other ends, *separation* or non-interference, and *prolongation*, the relation between two data which holds when one begins at the exact instant that the other ends. Moreover, between any two sense data *a* and *b* related either by separation, interference, or prolongation, I can detect an asymmetric order which I express by saying that *b* (for example) is *after a* or *follows* it, that *a*, on the contrary, is *before b*, or *precedes* it, either *completely* if *a* and *b* are separated (and *immediately* if *a* and *b* are in prolongation) or *partially* if *a* and *b* overlap. Let us call these order relations *complete succession*, *immediate succession* and *partial succession*. The last relation we shall consider is that of *simultaneity*, which connects two terms which begin and end together. This relation is of great importance both formally and theoretically.

These different relations clearly belong to the same family. In order to obtain some idea of their nature, we might begin by asking what this family relationship is.

The hypothesis of durations. Common sense, or rather, the spirit inherent in language, furnishes the following remarkable answer: all the immediate relations I call temporal are related in so far as each of them connects, not two of my sense terms, but two terms of another type altogether, namely two *durations*. (These durations are often themselves decomposed into *instants*. However, we shall ignore this second analysis which posits a second type of term whose instances are even farther removed from sense data than are durations). These new terms, the durations, are in turn linked to the sense terms by a relation *sui generis* which we shall call the *occupancy* of a duration by a term. Each of my sense terms thus has 'its' duration, and the temporal relations I apprehend between them are properly speaking relations between the *durations* they *occupy*. The relation of simultaneity therefore assumes a particularly important position. Indeed, it becomes the *identity* (and not merely the *equality*) of the durations of two data.

This conception may be called the doctrine of absolute sensible time. Between any two of my sense terms manifesting one or other of the so-called temporal relations, we interpolate two new terms, namely durations, which are to be the only real terms of the temporal relation considered. And this relation is extended to the sense data only through the agency of the secondary relation of occupancy which assigns a duration to each sense datum. Each of the temporal relations which I apprehend to hold between two *sense data a* and *b* thus ceases to be a *simple* relation proceeding directly from one to the other, and instead become a *composite* of three distinct relations, first linking the sense datum *a* to the duration α, then the duration α to the duration β, and finally the duration β to the sense datum *b*.

Let R denote a temporal relation between two sense data a, b; let ρ denote the corresponding relation between their durations α, β, in accordance with the above theory; and finally, let O denote the relation of occupancy of a sense datum to 'its' duration. With this notation the relation R may be expressed[1] as $O/\rho/\breve{O}$. (Fig. 3.)

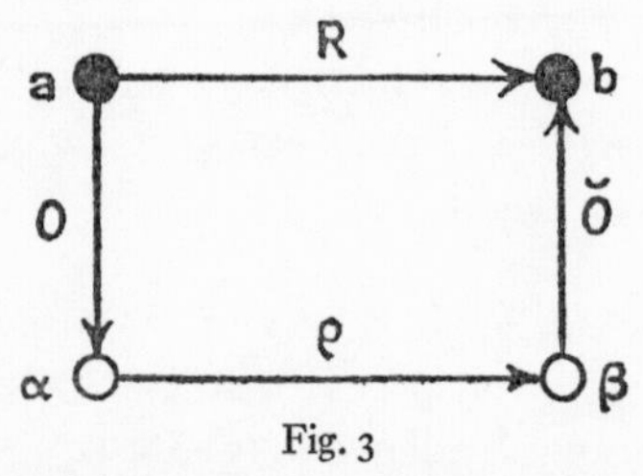

Fig. 3

This point of view has two advantages and one disadvantage.

In the first place, it provides a striking explanation of the family

(1) These logical symbols, in addition to others we may occasionally use, are borrowed from the *Principia Mathematica* of Whitehead and Russell.

relationship of the various temporal relations among my sense data: this family relationship consists in the fact that these relations admit *durations* as their immediate terms, and that they connect sense data only through the durations they occupy.

Moreover, the hypothesis of durations accounts for the particularly remarkable formal law according to which two simultaneous (i.e. entirely contemporaneous) sense data bear the same temporal relations to all other terms: they are simultaneous with, precede and follow precisely the same terms. Thus simultaneity faithfully *transmits* all temporal relations. With respect to any one of them, two simultaneous data are interchangeable; a set of simultaneous terms forms an absolute community; indeed no temporal relation distinguishes between them. These relations are attached, not to one or other of these terms, but to all of them quite indiscriminately. If I denote one of the temporal relations between my sense data by R, and the relation of simultaneity by S, then the 'transmissibility' of R by S may be written $R = S/R/S$ or schematically as in Fig. 4.

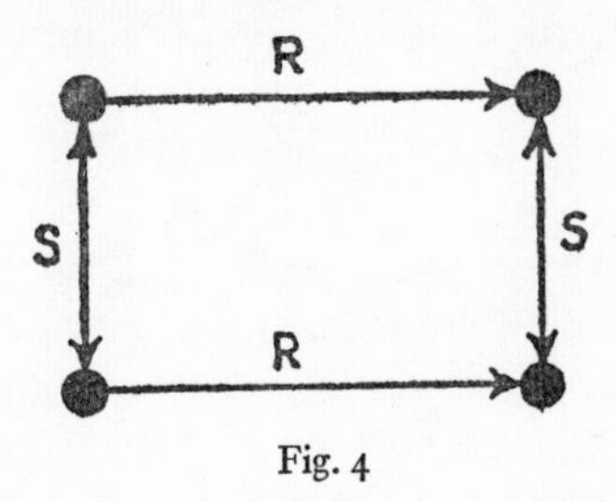

Fig. 4

Through this diagram we can see the relation R being transmitted from one pair of sense terms to the other by sliding in the direction S.

The hypothesis of durations makes this remarkable property of simultaneity appear to be analytic. Indeed, so far as this hypothesis is concerned, simultaneity means occupancy of the same duration; on the other hand, any temporal relation between sense data is actually a relation between the durations they occupy. The property in question may thus be reduced to the following: 'The relations my sense terms derive from the relations between the durations they occupy are identical for all of those terms occupying the same duration.' The durational hypothesis thus accounts for one important formal property of temporal relations, but only for this one. For example, it provides no explanation for the transitivity of succession or the symmetry of interference.

The drawback of the durational hypothesis is its prodigality with entities. When I say that a datum *b* – a certain thunder clap for instance – follows a datum *a* – for example, a certain flash of lightning – I am required by this hypothesis to have apprehended no less than four types of object: the two sense data *a* and *b*, the two durations corresponding to them, the occupancy relation which implements this correspondence, and finally the relation which results from the fact that the duration of *b* is *after* the duration of *a*. To think: 'That thunder clap followed that flash of lightning' would be to think the duration occupied by that thunder clap came after the duration occupied by that lightning flash. But isn't this exceedingly complicated? Even if I could be certain of apprehending all these elements – the flash of lightning and the thunder clap, their durations, the relation of each to its respective duration and of these latter to each other – the principle of economy, taken to be a methodological precept, would nevertheless prevent me from using this analysis of sensible temporal relations as a foundation for any structure which can be erected without its help, for example, all of the 'geometries of sensation' which appear in the last part of this work. It is good to avoid the superfluous assumption even of self-evident truths. But the durational hypothesis is entirely lacking in self-evidence. While I am certain of apprehending the lightning flash, the thunder clap and the sensible temporal relation which orders them, I am still in doubt with regard to these intermediate terms, the durations, and the occupancy relation which they would use to interpose themselves between the two primitive terms. Do they have truly simple natures, which I can understand, or are they, on the contrary, mere shadows of words? The principle of economy, now taken to be a probabilistic precept, invites me to forgo, if I can, such a costly hypothesis. It may be that my sense terms do not possess durations, and that consequently there is no relation between these terms as their durations. Perhaps there is nothing beyond the terms themselves and their direct temporal relations – simple and unanalyzable connections.

But what can I mean by the 'duration' of one of my sense data when I say that durations do not exist? This question must be answered.

Sensible duration conceived as a class of data. This duration is a 'non-entity', in other words, it is no longer the *simple* object it was before. It is now nothing more than the *class* of data simultaneous with the given datum, this datum itself included. (An analytic consequence of conceiving durations in this way is that a given duration is always the duration of some datum, whereas this conclusion could only be drawn synthetically under the hypothesis that durations are independent entities.) Depicting a sensible duration means thinking 'this, and all that is simultaneous with this'.

We cannot object that this is a vicious circle on the grounds that *simultaneous* means nothing except *occupying the same duration*, for this is precisely the analysis we are denying. The difference between the theory of duration-objects and that of duration-classes lies exactly in the fact that the former regards durations as simple and the temporal relations among sense data as complex, whereas the latter insists on the contrary that the relations are simple connections and the durations complex. Thus, if we say that the simultaneity of two data consists of the complex: *occupancy of the same duration*, then we are assuming the fundamental premise of the theory of duration-objects; in other words, we are postulating the existence of the very thing in doubt. Far from the theory of duration-classes containing a vicious circle, it is the objection itself which conceals a *petitio principii*.

The formulation of the concept of duration-classes is only the first example of a logical method[1] to which we shall remain constantly faithful. We shall also apply it to the extensions of sense terms and to their qualities. Since it will govern our treatment of the three important domains, space, time, and quality, it is worth our while to examine briefly the range and scope of the method.

The reduction of a sensible duration, an extension, or a quality to the class of data which 'have' or 'fill' this duration, extension, or quality is a kind of nominalism. But it is a nominalism which is at once more limited and precise than the traditional variety: first, because it recognizes the necessity of preserving a universal in the form of a *relation* which unites the members of these classes; and second, because it is not content with vaguely assigning the name

(1) Cf. B. Russell, *Principles of Mathematics*, Principle of Abstraction; *Our Knowledge of the External World*, passim.

resemblance to each of those constitutive class relations which act as replacements for ideas, giving the impression that it refers to a single relation which remains the same in all cases. Of course, the relations which hold among sense data 'of the same duration', 'of the same extension', 'of the same quality', are all of the general type *resemblance* which includes all symmetric transitive relations. However, it is clear that these three kinds of 'resemblance' must be different.

It is important to grasp the twofold significance of this method of logical construction. In one sense, it corresponds in each case to a particular hypothesis concerning the relative complexity of objects, as we have seen in connection with temporal relations. But in another sense it shuns all hypotheses, and therein lies its universal value.

For example, suppose I wish to study the temporal relations among my sense data. I am certain that these data are there. I am equally certain of apprehending these relations (without deciding whether they are simple or, on the contrary, whether they admit further analysis). When I say: 'That thunder clap followed that flash of lightning', I state a fact. Whether analyzable or not, it remains perfectly definite. Now, when confronted with such facts these *durations* soon slip into my language and into my very thoughts. But the fact that we find it convenient to employ them is compatible with a hypothesis that reduces them to logical constructs whose only elements are sense data and the temporal relations between them. Therefore, the wisest course is to understand them in this way. Indeed, this is the only course which risks nothing, for it is possible that these durations do not possess a more fundamental meaning. And even if they do, the artificial meanings we have ascribed them are for all that no less valid or less capable of the same usages than the simple meanings. All the propositions in which I have employed duration-classes to discuss the order of my sense terms remain true for these durations: they are in addition simply true of duration-objects.

In our geometries of sensation, we shall therefore only use duration-classes (and likewise extension-classes, quality-classes). But this entails nothing more than sticking closely to *sense data* and to the relations we apprehend among them. By using only these elements to define classes which provide all the services we require of durations, extensions, and qualities in the expression of facts, we do not exclude the hypothesis which says that these temporal, spatial, qualitative relations are complex connections which relate one sense term to another only through the agency of participation with non-sensible terms which would be the real durations, extensions, and qualities. For our purposes, this hypothesis is quite immaterial. In order to avoid either adopting or rejecting it, let us lay down the fundamental relations between sense terms without deciding whether or not they conceal complexities in which non-sensible terms might be lodged. This amounts to dealing with these relations *as if they were simple*, although in reality we decide nothing of the sort.

Concerning the reduction of temporal relations to a single one. Apart from the hypothesis that durations are simple objects, I can conceive another way of analyzing the temporal relations among my sense data. Without compelling them to pass through intermediate terms such as durations, I can investigate the possibility of reducing all these relations to a few, or perhaps to even one, of their number.

Take complete succession for example: all other immediate temporal relations invariably coincide with some logical composite of this one. By using the symbol '=' to signify *factual equivalence* as it does in mathematical logic, I can write

a precedes b = b follows a
a is during b = each x which precedes b precedes a and each x which follows b follows a

and so forth. Do these equivalences not define all the relations of sensible time in terms of the single relation *precedes*? Its mere feasibility makes such a reduction mandatory for the mathematician and geometer of time. All that he retains of a relation is the order it establishes among its terms: so far as he is concerned two inseparable relations are indistinguishable, for one merely duplicates the other without introducing a new order into the universe. His rule is not only to refrain from multiplying the number of entities, but actually to *reduce* their number as much as possible.

Right away we shall adopt the same rule. But at present I do not seek the most economical reconstruction of the order of my flux of sensation. I am simply inspecting the relations I discover there, among which it will immediately be necessary for me to choose the links of which I shall especially want to make use. I try to apprehend each one of them just as it presents itself to my mind. Now it seems rather doubtful to me whether my perception of the temporal relations among my sense data reduces basically to the perception of a single one of them, for example, complete succession. For some of these relations would *consist* of a complex nexus *involving the totality of my data*, past, present, and future. Thus, if I said that a datum *a* is during a datum *b*, I would mean that *a* follows all the data *b* follows, and precedes all the data *b* precedes; because this is the very meaning of the relation *during*. But could I ever *ascertain* a fact of this form? Undoubtedly, no. At best I might be able to infer it. Now it seems to me that *a during b* is a relation I can sometimes directly detect between two sense terms. Hence, it is very likely that what I detect between *a* and *b* is the presence of a relation which involves no other sense terms – this would be precisely the relation *during* – and that only later do I infer this second relation between *a* and *b* which involves all my other sense data. Without a doubt this is an instance of the general principle that a definition cannot possibly express the true meaning of a sometimes verifiable relation which it recasts as a complex relation involving all the members of an infinite class.

A second difficulty arises from the necessity of making an arbitrary choice. Of the two relations *follows* and *precedes* for instance, which is the original and which the logical inverse? This problem is easily solved by the geometer of time, but it stops the philosopher in his tracks. This problem, which applies equally to all asymmetric relations, may moreover be purely linguistic. But the first problem is only too real. It therefore seems that the temporal relations among my sense data are composed of not one, but several basic relations, and intuition appears to confirm this conjecture.

Definition of a natural family of sensible relations. This brings us back to our original question: what is the nature of this obvious family resemblance uniting what I call the temporal relations among my sense terms? For I have just admitted that there are *several* basic relations. The hypothesis of real durations implies that this family resemblance resides in a common reference to durations. But, in the hypothesis which posits the durations as complex and the temporal relations among my data as simple, are we not obliged to say that this family resemblance is ultimate and incapable of analysis? These relations would, therefore, all have an indefinable *temporal aspect*, which would be the irreducible residue of the general idea of sensible time.

This hypothesis of a simple temporal quality is not absolutely indispensable, however. Indeed, certain formal laws establish connections among these relations which perhaps express all their affinity. First of all there is the law of transmission with respect to simultaneity, which, as we know, gathers all temporal relations together around it. Then there is the law which says that any two of my sense terms are always connected by one or other of these relations. Thus we no longer have a regularity in virtue of which certain relations are inseparable; on the contrary, we see them separate, but only in order to share the mass of my sense data, in the same way as several sportsmen divide the territory of the beat among themselves.

Now may I not claim with good reason that relations, the sum of which covers the whole of my sensible process, capture the same aspect of the whole universe, that they form a network covering the same facet of experience?

While the current conception extracts a *duration* from each of my sense data (and then decomposes this duration into *instants*), we have seen that the only thing beyond dispute is my apprehension of certain relations among these data which I call temporal. It remains doubtful whether these relations involve durations, or *a fortiori*, instants, or even that they all have a common property.

On the nature of the laws of sensible time and related laws. In connection with temporal relations, let us say a few words about the character of the laws expressing the general properties of relations among my sense terms. For example, simultaneity transmits each of these relations; any two of my sense terms are connected by one of them; temporal inclusion is transitive; complete succession is transitive and asymmetric; simultaneity is transitive and symmetric. How do we discover laws of this sort? What is the degree and nature of their perspicuity?

To a large extent the question is governed by the opinions adopted concerning the simplicity or complexity of each of the relations considered. If I posit real durations, then the properties of simultaneity become analytic; if I make a particular relation a composite of certain others, then the properties of the first derive logically from the properties of the others. In general, the result of assuming complexity in a system of notions is that one or several of its properties become analytic. However, there are always some synthetic properties present and, even if all temporal relations were to be reduced to a single one, the properties of the former could not be determined thereby. The greater or lesser number of laws depends on the view I take regarding the content of these various relations; however, a nucleus of formal properties which cannot be reduced to identities is present in every case.

Is there an *a priori* 'chronology', as Kant thought, or do the axioms of this science of sensible time only possess a self-evidence similar to that of the incompatibility between black and white, silence and noise, the appearance of squareness and the appearance of roundness? Doubtless this self-evidence is not of a purely inductive nature, and rightly holds a sway over the imagination, but is not essentially different from that which links pleasure or pain to such and such a sensible object. This is a delicate question, and we can do no more than draw attention to it.

CHAPTER III

Global Resemblance

Sometimes, when I hear a sound, scent a fragrance, savour a taste, I recognize its quality and say that it is similar to a sound, fragrance, or taste of which I have retained an image. This resemblance between two sense terms has degrees like all resemblances, but it is not restricted expressly to any one of the various aspects of the terms it relates. Since it treats each of them as a single whole, we shall call it *global resemblance*. Later on, we will discuss two more specialized kinds of resemblance.

Occasionally we think that two data never resemble each other in all their aspects, since they are apprehended in two different total perceptions; and even if all the circumstances were reproduced, I would nevertheless feel that they were not the same as before, since they would no longer be new to me. On the other hand, any two data always resemble each other in *some* respects. Thus the global resemblances of my sense data would merely form a chaos.

However, we are confusing two things here. A resemblance can be either *direct* or *indirect*; from my point of view it either proceeds directly from one of its terms to the other, or else consists of a *common relation with some extraneous term*. Now, we speak of nothing but *direct* resemblances here while it is only through *indirect* resemblances that anything whatever resembles, or differs from, anything else. The sense data which accompany a certain datum *a*, even to the extent of covering and surrounding it, the images which arise from it, the feelings it stirs, and all these taken as a single whole, are just so many terms logically unrelated to the datum *a* and hence also to the direct resemblances it may or may not have to another datum.

But let us even agree to include as part of the quality of a sense term everything it happens to accompany and all it evokes in my mind. Let us admit that the opposite feelings engendered at different times by the same scent, the various sights and sounds it conjures up, are incorporated in the olfactory datum just as inseparably as is the quality in its subject. We must confess that this quality has two parts, one which is indefinite and nebulous, full of reverberations and echoes, and a core of constitutive quality. I perceive this division in the contrast which makes me say: 'Nothing has changed since yesterday, and yet everything seems different to me.' *Nothing has changed :* I recognize today's sense terms to be the same as yesterday's through an essential quality quite removed from the more extensive one which involves all aspects and relations. *And yet everything seems different.* I notice a striking contrast between the data of today and yesterday, but which does not apply to the central quality.

Of course, I might falter and even make a mistake here. But we ought not to require the mind to function smoothly and flawlessly in order for us to recognize the existence of its power. Hesitation and the danger of error are everywhere. No judgement is infallible, but the possibility of error alone does not make it futile. Our attempt to extract the constitutive quality of a sense datum from the cloud of associations and feelings which surround it might end in failure. But it is by no means an absurd thing to attempt. A wine taster must admittedly make an effort when passing judgement on the identity of two flavours separated by an hour of life and by a complete alteration in his state of mind. It is an effort of attention, of abstraction, if you please, but nonetheless a real effort, one whose purpose is not illusory.

The three logical forms of resemblance. Global resemblance has an additional feature: it is a relation which possesses *degrees*. We must not confuse this type of 'more or less' in our certainty of its presence. This second kind of gradation is universal and attaches itself to all relations quite indiscriminately. Take, for example, a relation R. There may exist three terms *a*, *b*, *c*, such that *a*R*b* is more certain to me than *a*R*c*, so that *a* bears the relation R more certainly to *b* than to *c*. But this does not indicate a gradation in the relation R itself; it is nothing more than the degrees of the two assertions *a*R*b*, *a*R*c*, each one of which has its meaning and degree of evidence independently of the other. On the other hand, the proposition '*a* resembles *b* more than *c*' does not effect a fusion of the two independent propositions '*a* resembles *b*' and '*a* resembles *c*', but links up the three terms directly with a simple self-evidence that does not consist in the comparison of two self-evident truths.

This amounts to saying that a relation which admits degrees is a three-termed relation, for example: '*a* resembles *b* more than *c*'. Let us call it the *order of global resemblance*. We cannot doubt the reality of this relation. I sometimes very clearly apprehend this relation among three very neighbouring and yet patently different sense data, for example, three sounds, odours, or shades. We must be careful not to postulate that among any three data, even when they belong to the same sense, there is always one which lies between the two others, for this is very doubtful.

We are, therefore, confronted with two relations, one of two terms, *global resemblance*, and the other of three terms, *order of global resemblance*. It might be possible to add a second two-termed relation, namely, the *perfect global resemblance*, which I apprehend between two data, one of which appears to be an exact duplicate of the other. However, we can construct a general hypothesis which effects a reduction of both simple and perfect resemblance to degree of resemblance. '*a* resembles *b*' would then mean 'there is a term *x* such that *a* resembles *x* less than it does *b*'. '*a* resembles *b* perfectly' would mean on the other hand 'there is no term *x* which *a* resembles more than it does *b*'. There would be a resemblance between *b* and *a* just in case *b* resembled *a* more than it did some other term; there would be perfect resemblance between *b* and *a* if and only if *b* resembled *a* at least as much as any other term does.

It is true that this last formula clashes with the principle which claims that no verifiable relation can involve the totality of my sense terms. But am I ever certain that two terms resemble each other perfectly? Thus, it is possible that perfect resemblance is never recognized directly, but only inferred from the fact that my imagination furnishes no intermediate term. If this is the case, the objection vanishes.

In view of the fact that simple resemblance can have a content as near to zero as we wish it, and that perfect resemblance always remains uncertain – the one lost in vagueness, the other in the realm of the ideal – degree of resemblance is from any viewpoint the most positive relation of the three.

CHAPTER IV

Qualitative Resemblance and Local Resemblance

Beneath the surface of global resemblance I can discern certain *partial resemblances*. The ability to do this is something new, for it could be supposed that I discern with acuteness of hearing whether or not a sound *b* is a perfect reproduction of a sound *a* without being able to ascertain, of two sounds which differ, whether they differ in intensity, tone, or duration, and on the contrary resemble each other in pitch. I would not be able to recognize a note unless it is played in the same way on the same instrument. Likewise, I would not be able to recognize a colour, a form, a region of my visual field individually, but only in conjunction with one another; and the same goes for the sense of touch. There would thus be only one, and not several ways for my sense terms to resemble each other (directly, of course).

But the situation is actually quite different, at least for certain senses. So far as tactile impressions are concerned, and this applies more strongly still to visual impressions, global resemblance divides into two kinds of partial resemblance, which we shall call *local* and *qualitative resemblance*. This division is highly important. I am motionless; two identical sparks flash in succession at the same point. The two data I have just apprehended resemble each other in every respect globally, like two flavours or two smells. Now suppose one of the sparks occurs on my right, the other on my left; they still resemble each other, although not so strongly as before. This is not the whole story: they resemble each other acutely in one respect, and not at all in another. Consider two sparks which differ in colour, but not in position. As in the preceding case, they resemble each other somewhat; and just as before there is a strong resemblance in one respect (that in which the two preceding sparks *did not* resemble each other at all) and not the slightest resemblance in another respect (precisely that in which the two preceding sparks were the same). Let us call these two kinds of partial resemblance between visual terms *local* and *qualitative resemblance* (in the narrow sense: for, taken literally, the quality of a sense datum doubtless includes its immediate locality).

The same distinction can be discovered in the sense of touch, and perhaps even in all senses. Indeed, it is possible that the immediate locality of a sense term is always distinguishable from the rest of its quality. But in that case the above distinction would assume in the order of experience only in the case of touch and sight. For even if we allow that an odour, for example, has just as much claim to have an immediate locality as does a visual or tactile datum, this locality is still *the same* for all odours. Each olfactory datum permeates the whole of the olfactory field. Locally, they all resemble each other: local resemblance induces *no classification*, no order amongst them whatsoever, and consequently cannot enter into the expression of any law governing my olfactory universe.

Again, let us consider the kinesthetic sense. To each of its data let us assign a local character which is distinct from the mass of its quality. Now, we are no longer dealing with terms which all resemble each other locally. Indeed, any local resemblance does no more than duplicate their global resemblance: for two kinesthetic data to possess perfect local resemblance, it seems necessary that they be completely identical and derive from the same parts of the body, which must be placed or displaced in the same manner. If this is the case, the perception of local resemblance and diversity is, once again, fruitless. Before, it failed to establish any order among the olfactory data because it could not distinguish them. Now, it plays a dual role within the apprehension of global resemblances, uniting those terms which the latter unites and separating those it separates.

From the viewpoint of the order of my sense data, the co-existence of local and qualitative resemblance is of no importance as long as these two resemblances are on an equal footing *or* one of them links any two terms whatsoever. This does not in any way enrich he framework of possible laws. The important thing is the existence of distinct local and qualitative resemblances, in other words, the presence of *two networks of resemblances which intersect*, and which classify the same data in two different ways. This richness of structure is possessed by sight and touch alone. This is why they yield especially interesting geometries. More meagre ones exist, however: we shall see that it is possible to distinguish one, and even several kinds of empirical geometry which use global resemblance only.

Are global, qualitative, and local resemblance elementary or complex relations? Do they proceed directly from one sense term to another, or do they consist, on the contrary, of the participation of these two terms in a common entity which would be their total quality, their quality in the narrow sense, their immediate locality? We have indicated the general gist of our answer in the case of temporal relations. On the one hand, it seems doubtful to us whether the resemblances in question actually involve these special entities, as well as the auxiliary relations by which sense terms partake of them. On the other hand, this hypothesis has no relevance to the problem which occupies our minds and which concerns not the content of the relations among my sense data, but the ordered network they delineate. We must, therefore, remain neutral. To make no assumptions about the simplicity or complexity of a given relation amounts to treating it as if it were simple.

Like all resemblances, qualitative and local resemblance allow of degrees and a maximum. Both can, therefore, be divided into three relations: a pure and simple resemblance, a perfect resemblance, and a degree of resemblance among these terms. We have already considered this triad of relations in connection with global resemblance, and we have investigated the possibility of deriving two of its members from the third.

CHAPTER V

Relations of the Local Resemblance Family

Local resemblance acts as a kind of centre for a family of relations which cluster round it. This family consists of all those relations which *are transmitted* by local resemblance in the same way as temporal relations are transmitted by simultaneity, that is, all relations R which satisfy R = L/R/L, where L stands for perfect local resemblance.

First examples are *inclusion*, *overlapping*, and local *separation* which recall the corresponding temporal relations.

The relations of position. But the local resemblance family also contains relations without temporal analogues. One evening I gaze down from the top of a mountain at the lights of a town, and each gas-jet appears as a brilliant point. Do I not notice a definite resemblance among every triad of glittering points which derive from three lamps in a straight line, no matter where they are situated in my visual field? It seems so, and it is this relation among three visible points, or, if you like, this special resemblance between two triads of visible points, that we wish to consider. Let us call it *alignment*. (We do not determine here whether this relation only holds among three data deriving from three objects in a *straight line*, or if, in addition, it is present among three data whose corresponding physical objects are simply *in the same plane as my viewpoint*. Both hypotheses will be discussed later.) It is transmitted by local resemblance: three visual data bearing perfect local resemblance to three other data in alignment are themselves in alignment. Three sparks, each of which is locally perfectly similar to one of three preceding sparks which appeared to be in alignment appear aligned in their turn. It is easy to imagine other relations in the same family. For example, I may perhaps apprehend a particular resemblance among all the pairs of visible points deriving from pairs of small objects separated by equal visual angle: we may name this resemblance *equality of separation*. Let us term *relations of position* those relations of the local resemblance family which do not have analogues in the family of temporal relations.

As first examples I chose very small sense data (sparks, distant lights) because it is customary to conceive of the relations of position as illustrative of the geometric relations among points. But there is no doubt that they correspond rather to the relations among volumes, which, as we have seen, also provide a foundation for geometry. Moreover, they can undoubtedly hold among data of any magnitude whatever. For example, the real sensible relation of alignment is that which leads us to say, of three large visual terms, that 'there is one straight line which crosses them'.

Relations of position really appear more suitable than any others so far considered for illustrating the abstract structure we call geometry. So much so that if I search for illustrations of this structure in the networks formed by other relations among my data, I shall perhaps be thought incredibly blind in seeking these same relations of position in places they are not. But this criticism results from not conceiving geometry abstractly enough.

Later on we shall see that these relations are not essential to the order through which sensible nature confirms the laws of science. Indeed, they are inseparable from the complex connections formed by relations in other families. In the framework of the world, they therefore duplicate these connections; they may perhaps even be reducible to each other, and possess no basic content whatsoever (cf. Part Three, Chapter VIII).

*

Such are the elementary terms and relations of the sensible process. We have merely sought to present them before the mind's eye. If we have examined various opinions about their nature, it was only to give them a more distinct appearance for the purposes of discussion. Hence our sole objective was an unsystematic perception; for it is very important to apprehend these elements of facts as facts in their own right, quite outside any theory.

Perhaps we have given adequate proof that no hypothesis has been adopted in regard to the representative content of the sensible relations just enumerated. The simple distinctions between them should suffice for our purposes as long as what interests us is the order they impose on nature. Likewise, we have made no assumptions about their conditions or histories. When I am considering the immediate reality which I indicate to myself by thinking 'this relation', I know nothing whatever about distant realities such as the body, the sense organs, and the nervous system. However, in order to describe this relation to you, I have to say: What you apprehend in such and such a situation involving various objects and your body. But this detour is only provided to make you think of this relation, and I am really speaking of this alone. Finally, I do not know the history of these relations which present themselves to me as the elementary links of nature. I do not claim that they have no history. I do not know what is known by an animal, a savage, or a child. I do not think that my perception has always been what it is now. What I am considering is the sensible universe of an adult, or even an adult physicist. It is here that we must discover how physics is interpreted.(1)

(1) One may perhaps be surprised not to find in the list of ordering elements of sensible nature any mention of a primitive quality of *spatiality* (voluminosity, extension) which would be common to all data. But a property common to the members of a set does not impose any order on them. We would be forgetting this quite simple logical truth if we were to see in the spatiality of sense data the quality of geometricity (if it is permitted to express a meaningless statement by a barbarism) which itself submits them to geometry. If among the authors who have written on 'space', some have thought that an analysis of the geometric order of experience should furnish an analysis of its intrinsic spatial quality, then they were assuredly deceiving themselves, and we shall not embark on this chimerical operation. But if others have been able to believe that the simple indication of this indefinable quality of all sensible experience either established, constituted, or replaced the analysis of geometric order it provides, they were no less mistaken. The last part of this work therefore cannot be criticized for unravelling with difficulty a knot which is easily cut by intuition or words.

Part Three

SOME SENSIBLE GEOMETRIES

Introduction

We have studied the formal structure of geometry. In addition, we have drawn up a list of the primitive relations that each of us apprehends in sensible reality, simple connections constituting the whole order that it comprises, elements of all the texts we shall ever read in it. Let us now turn to the reading itself and rediscover geometry in the book of nature. But the task is too great to be accomplished at the first attempt. Indeed, geometry does not enter nature before physics, but actually through physics itself, from which it merely extracts the most general outline. It has already been revealed that experience rests not on space, but on bodies, or more generally on the sensible. However, geometry insinuates itself into the expression of any experience through the situations of the objects and observers, which are part of the circumstances of any sensible fact. Geometry exists in these expressions alone; it cannot be isolated from them. Its sensible truth is none other than that of the set of physical statements which contain it.

But any statement of physics contains some geometry; and the sensible effects of various branches of physics (astronomy and optics, for example) are intermingled, and fused with these effects is that resulting from the physiology of the senses. Far from amounting to a few simple isolable facts, the geometric order of my experience has for its content all my knowledge of the sensible world in an inseparable unity. This universality is the essence of the problem. But it confuses the mind at first.

Let us then begin with problems of the same type, but commensurate with our mediocre analytic faculty, by placing ourselves in a *simplified experience*. Since the geometric order of the sensible world resides in the set of its laws, let us begin by imagining worlds whose laws are simple enough to be apprehended in a single glance. In this way we shall make our first contact, as it were lightheartedly, with the problem of the application of geometry to nature. And we shall accustom our mind to some of its larger aspects.

We shall thus determine an experience by fixing its conditions arbitrarily. For example, we shall imagine a being reduced to the sense of sight moved about in an unchanging universe without receiving any information except from the change in scene. In this universe we shall place only the objects we choose, and distribute them as we wish. If we so desire, the subject will have a body invisible to himself. Or else we shall consider a being who experiences nothing apart from the contractions of his body. We shall place him in a universe devoid of any perceptible differences, and there we shall watch him wander about; but every time his movements bring him back to a certain position, we shall inform him of his return by a particular impression.

We shall assume the still more subtle faculty of understanding the following scientific constructions: homogeneous media, material points, luminous points, and of delineating the corresponding perceptions. We shall imagine senses of infinite delicacy, a kinesthetic sense for which no two sensations are identical unless they betoken two exactly similar movements or attitudes, and a sense of vision which distinguishes two different luminous points no matter how small the angle between them. This perceptive sense will be perfect not only in its subtlety but also in scope. A vision embracing the whole of space in a single glance, and even a vision, if this term remains permissible, which in one single act distinctly apprehends all material points of a space completely filled with them – we shall allow ourselves these fictions, so as later to study them in the most serious way possible.

Are we not invited to proceed in this way by the very science whose sensible range we are investigating? Physics does not hesitate to begin its approach to the real through the fabrication of entirely idealized simples: the material point, the luminous point – could there be a freer wave of the wand? And, having adopted these as objects, it sets us an example by studying them in the greatest possible detail. It certainly knows that these idealized objects cannot be found in nature. But it does not see this as an excuse for haste without rigour. It fortifies itself by mastering these schematic universes; then it voluntarily relinquishes some of its fictions in order to let the difficulties which draw it closer to nature enter one by one. The physicist does not dive into the ocean of reality without taking precautions, for he would run the risk of drowning. Supported at first by a lifebelt of ideal simplifications, he removes it bit by bit as he progresses in skill. Let us imitate him; for our study follows his, and is fundamentally nothing more than its analysis.

The mind cannot do without these gradual approaches. Of the few authors who have considered the empirical meaning of geometry or, more generally, the sensible content of physical facts, there is not one who has not placed himself in experiential conditions which are schematic in the extreme. But since the mind is quick to follow them in this field, they have allowed their assumptions to remain implicit, or have at most indicated them by a quick word or two. To insist would have seemed pedantic: is it not the philosopher's privilege to discard detail in order to arrive at a concise viewpoint by the shortest possible route? Thus, in allowing their simplifications to remain shadowy and thereby renouncing any rigorous construction, they only aim to expose the mind to a vague idea of an empirical nature whose message remains nebulous, for while their failure to relate it distinctly to the hypotheses involved doubtless succeeds in veiling the gap which separates it from reality, it also deprives them of the ability to measure and reduce it.

Therefore, between this 'natural' method of analyzing sensible order and the apparently artificial method we prefer, lies only the difference between the nebulous and the distinct. Plainly postulating conditions which are to a large extent fictitious, we shall study the results objectively. It is not that we are lured by a spirit of unreality. On the contrary, the precise list of assumptions which lie at the bottom of such schemas enables us both to take them at their true value and subsequently to improve them. For the spirit of approximation must not manifest itself in philosophy as a spirit of negligence. It does not consist in being lavish with the 'close enoughs' and the unambitious conclusions. Its aim is not to disarm criticism, but instead to expedite it. It states its assumptions plainly, it shows clearly what results from them. Far from hiding the artificial aspect of its work, it aims to exercise the mind without deception, in order to prepare it for a better approximation.

As soon as we cease to regard the light ray as a straight line without breadth in order to pave the way for the wave theory of light, the whole of optics has to be rebuilt on this new foundation. Nothing remains of the previous structure apart from a recollection which guides the reconstruction by whispering what features, roughly, ought to be rediscovered in it. Likewise, the sensible geometries which follow are neither parts, nor foundations, nor origins of the almost infinitely more complex geometric order of our world. Erected on fictions, they can be nothing but useful approximations, like ray optics or the mechanics of the material point.

The scientific turn of mind we shall lavish on our imaginary observers, while impoverishing their experience, is really an expository device. Our investigation is not in fact psychological. In each of the worlds we posit, we examine the order actually present and not the portion that would be discovered by a determined use of intelligence, memory, and curiosity. Our subjects enjoy infinite intellectual power in order that they may discover all the order there is to be discovered. With regard to the structure of the sensible world, their intelligence performs the services of a perfect guide in a narrative whose real object is purely descriptive.

Let us state the problem as precisely as possible.

Consider any geometry, for example that in which the primitive terms are *points* and the sole primitive relation is *congruence* of two pairs of points. Let $G(p,C)$ be the system's set of axioms, expressed in terms of a class p and a relation C between two pairs of members of p. Moreover, let s be the class of my sense terms; let $R_1, R_2, \ldots, R_n$ list the various relations I observe among them, and let $E(s,R_1, R_2,\ldots,R_n)$ be the set of laws which I shall be led to regard as inductively probable. Discovering an illustration, a 'solution' of the formal system $G(p,C)$ in sensible nature means constructing, in a logical manner using the relations R and the class s, a relation C_o and a class p_o such that $G(p_o, C_o)$ is entailed in $E(s,R_1,R_2,\ldots,R_n)$.

Let us illustrate this with an example. We have all seen as children those drawings which depict something we cannot distinguish at first sight, where the idea is to discern a giraffe or a lion in the contours of a landscape which at first sight seems deserted. When we have 'discovered' them, we have seen nothing new. The outline of this hillock was the lion's rump, and the knot in this tree trunk its eye. Into this network of lines we had read a certain structure, namely the landscape, and now we have just read a second structure into it, namely the lion. As for the lines themselves and the elementary relations – angles, distances, intersections – which in the final analysis determine the whole drawing, we have in these the substance of the remainder, the arabesque itself, into which we can first of all read a landscape by noticing that its elements manifest a certain order when grouped in a certain way, and then a lion by observing that a different grouping brings to light a second structure. The drawing I have before me is sensible nature. The elementary links which I know how to spell out, so to speak, are the primitive relations among my data. The form I attempt to read into it is, for example, the geometry $G(p,C)$. What groups, taken as elements, make this structure G appear in the relations which stem from their composition? Could there even be several methods of grouping which satisfy this requirement – might there even be more than one way of finding a lion in the landscape?

In general, the relation C_o and the class p_o will be complex: the relation C_o will be a logical composite of $R_1, R_2, \ldots, R_n$ and the class p_o will have as members, not members of the class *s*, but *classes* (if not even classes of classes) of these members, defined by means of the relations $R_1, R_2, \ldots, R_n$. However, it might happen that one of these relations R together with the class *s* already form suitable meanings for C and *p*. This is what would occur if I saw every point of space and apprehended their congruences directly – as we shall later on assume in one of our examples. But, *even then*, I would have to investigate whether the ordered network of my experience $E(s, R_1, R_2, \ldots, R_n)$ does not contain in addition *other* solutions of the geometry G, *other* meanings for its relation C and its class *p* – *complex* meanings – alongside this first system of simple meanings.

This logical formation of relations and terms possessing certain characters by means of relations and terms which do not possess them, this definition of 'points' from that which is not a point, of 'congruence' in terms of relations not one of which has the properties of congruence, may indeed seem suspicious. We are tempted to think that if geometry is not in the realm of the simple, it cannot suddenly materialize through synthesis.

But nothing is more commonplace than the creation of new formal properties by simple logical combination. The relation $>$ 'greater than' is *asymmetric* and *transitive*: it follows that its logical inverse $<$ 'smaller than' is likewise, and that, on the contrary, the logical sum of this relation and its inverse $\lessgtr$ 'larger or smaller than' is a *symmetric* and *non-transitive* relation. At the very beginning of this work we came across our first interpretation of geometry in arithmetic: numbers grouped in threes filled the role of points and congruence was rendered by a group of two equations made up of various relations which reduce to addition and multiplication. Thus geometry has already made one appearance in a structure of terms and relations in which it was not elementary by means of logical construction alone. Nothing stops the situation from being the same in sensible nature.

In the sensible worlds we are going to imagine, the ordered network will, naturally, be composed of the relations whose presence in immediate reality I have recognized in Part Two. But in these simplified experiences, the only ones about which we can possibly make assertions with certainty in the present state of analysis, geometry is illustrated by structures which do not bring all these relations into play at once. There are *several groups* of relations each of which form one or several geometric networks. Thus the most remarkable feature of a 'geometry of sensations', more remarkable even than the sense or senses in which it takes its *terms*, is the group of directly apprehended *relations* it employs.

The spatio-temporal relations independent of the distinction between duration and extension do not yet play a role here, although they are undoubtedly of great importance in a less schematic view of things. But the relations of time, global resemblance, local resemblance, qualitative resemblance and the relations of position of sense data form, in an ideally simple experience, three types of geometric structure. One is in the combinations of *succession* and *global resemblance*. The second resides directly in the *relations of position*, if we take these as basic. The third consists in *simultaneities* and *local* and *qualitative resemblances*.

We shall first of all investigate the geometries of succession and of global resemblance of sense data.

These geometries split into two types, according to whether we consider only the data of an external sense, or adjoin to them kinesthetic data. It is true that the first type includes only rudimentary structures restricted by the propositions of *analysis situs*. And it is with this type that we shall begin our discussion.

CHAPTER I

Succession and Global Resemblance

(data of any external sense)

Let us restrict ourselves first of all to the case of a geometry reduced to the properties of linear order. In doing this we again raise, although in a completely different spirit, a little problem formulated by M. Bergson in the following terms: '. . . Imagine an indefinite straight line and on this line a moving material point A. If this point were to become conscious of itself, it would be aware of change since it is in motion; it would perceive a succession; but would this succession assume for it the form of a straight line?' (*Données immédiates*, p. 78).

What is 'the form of a line'? The question is not in fact the same for M. Bergson as for us. What does the 'form of a line' mean to us? It is a property as abstract as it is complex. We call *line* a class of terms linked by a relation obeying the axioms of linear *analysis situs*. What terms, what relations, are going to satisfy this definition, we have no idea at all; or at least, if we do not wish to feign such complete ignorance, if we do not consent to overlook a certain image which springs to the mind spontaneously at the very mention of the word 'line', we must realize that this image is but the first example of what we are after, and that the other examples, if they exist, are not obliged to resemble it. Therefore we cannot in any way restrict the nature of the elements and relations which form a line in the analytic sense; in sensible nature we are searching for observance of the axioms of linear order, and not a certain aspect which would be suitable only for a line.

M. Bergson, on the contrary, seeks nothing else. He does not say what he means by a *line*, but his answer to the question he poses gives sufficient indication of the meaning he assigns the term. The moving point's experience would, he says, assume the form of a line 'on condition that it could in some manner rise above the line it traverses, and apprehend simultaneously several juxtaposed points'. Whence this assertion? It arises from the fact that M. Bergson conjures up a peremptory image of what a line ought to look like. He imagines a set of simultaneous elements offering a certain primitive order which he calls juxtaposition and, seeing that this is indeed a line which satisfies the geometer, he does not dream that there may still be some, which are altogether different, fashioned from other elements and relations. The form of a line means to him this particular appearance. Having asked himself if the succession his conscious moving point experiences would assume the form of a line from its viewpoint, he then answers: undoubtedly yes, provided that it has this aspect, or assumes it through some illusion. But this is to abandon thought for imagination, to slip into the arbitrary. For us, the form of a line will be solely the laws governing lines: and if, as we shall see, these laws can be found in the intuitive order of an instantaneous apprehension, they can also be found in other aspects of experience.

True, M. Bergson considers that, wherever they may be found, these laws can only be thought of by means of this particular image of a simultaneous multiplicity he calls the idea of space; and he appears to think that, conversely, a being endowed with this idea and using it in his thoughts as a kind of blackboard cannot fail to transmit the intuitive order it presents him to everything he conceives with its aid. Thus, the application of geometry to nature would be entirely included in this single representation. But the first of these assertions does not concern sensible laws, but only the mental means required for thinking about them. As for the second, according to which the use of these means would determine the laws themselves, we shall see that it is untenable (Part Three, Chapter V).

Let us return to our problem, which we take in the general sense just specified. The moving point which describes straight line A 'is aware of change since it is in motion'. But this can occur in at least three ways. A horse pulling a carriage without seeing anything is aware of change in that it experiences the deformations of its limbs; the coachman, on the other hand, is aware of change in that he experiences the continuous flight of the landscape; finally, the passenger who casts a glance through the carriage window from time to time is aware of change in as much as he embraces a new scene on each occasion. Which of these three modes of experiencing displacement shall we choose? The horse's experience is formed by internal data which we are at present ignoring. On the other hand, the passenger's experience, which consists of impressions received in places having no systematic proximity relationships with each other, is too discontinuous to be included in a discussion confined to the relations of succession and global resemblance. Let us therefore posit an experience of the same type as the coachman's, in which the displacement is expressed without a break by a series of external data; and since we are restricting ourselves here to the two relations of succession and global resemblance, we shall single out the relatively undifferentiated sense of hearing in order to fix our ideas.

The simple open line. Let us, therefore, imagine a creature with only the sense of hearing transported along a line divided into small segments such that with each transit over any segment, say A, a sound of a particular quality *a* is heard. (In this situation we exclude the existence of two segments *which are indiscernible* by the sounds associated with them: this simplification is convenient, but we shall see that it is not, as we might believe, essential.) This path may be compared with the keyboard of an organ whose keys would be sounded by our subject in passing over them.

We assume that our subject conceives only two questions in connection with the sounds produced which constitute his whole perception. (We shall always understand by *a sound* an individual sound, and not an audible quality.) The first is: *Was the sound y after the sound x*? Let us try to find out if the sensible world so defined possesses laws, and if the subject recognizes a physics.

This physics can contain only two notions, namely the relations of succession and resemblance. (By resemblance we shall understand perfect global resemblance, and by succession, complete succession.) Their network is the whole structure of the universe. But, prior to the laws which combine them, each of the two possesses its own laws.

There are three laws of succession: (1) *the sound y follows the sound x* precludes *the sound x follows the sound y*; (2) *the sound y follows the sound x* and *the sound z follows the sound y* implies *the sound z follows the sound x*; (3) either *the sound y follows the sound x* or *the sound x follows the sound y*. Logic would express these laws by saying that the succession of sounds is an asymmetric, transitive, and connected relation, or, in one word, a *series*.[1]

(1) Cf. B. Russell, *Introduction to Mathematical Philosophy*, Chap. IV.

Now the series is one of the two forms which define the order of a simple open line: the succession of sounds is, therefore, a simple open line.

But this is still no more than the study of the succession of sounds, and not yet of the resemblances among them, of the flight of things, and not of their return. Nothing in it yet expresses the fact that the melody which comprises this sensible universe denotes what we call a displacement, and still less a displacement along a path of a certain sort. Does such an expression exist? This depends on the journey we make a moving object undergo. As long as we do not change its direction, the sounds heard are all different. The observable order is reduced to the monotonous pattern woven by succession. The fact that the sounds that succeed each other reflect what we call a progression through space is still unexpressed.

But as soon as we make the moving object turn back, its science develops two new branches. Beside succession a second simple quality appears, that of resemblance. Like the other it has its own laws which make it a symmetric, transitive, and non-connected relation. To the science of succession is thus added the science of resemblance. But in addition, a new complication arises.

Succession arranges all the sounds into a single series; resemblance, on the other hand, constructs classes of sounds, resembling each other, but different from all the rest. But the members of these classes of similar sounds are dispersed in the order of time: *the twofold structure which results from this fact is perhaps the most fundamental feature of sensible nature.* This mixture of successions and resemblances forms the whole of physics. In our subject's universe it admits a very simple law. Let us first ascertain what it is for us, and then what it is for him.

Of any three sections of a simple open line, there is one which is found to be *between* the two others and which we are obliged to cross every time we proceed from one of these two to the other. Now all the transits of the moving object through a given section are signalized by similar sounds. Thus, if A, B, and C denote three classes of similar sounds, one of these classes contains a sound in any series of sounds which come after a sound of the second and before a sound of the third or vice versa: this is the principle of interlacing successions and resemblances in our subject's universe. The principle is clearly quite empirical for him, without a shadow of intrinsic evidence or of necessity. But let us see exactly how it manifests itself in experience and what complex notions its formulation reveals.

In the first instance, it is the relation of a sound to two other sounds, one of which has preceded and the other followed it. Let us say that the first sound *separates* the two others. Secondly, it is the class of all sounds similar to a given sound: let us call this class the *note* of this sound. Finally, it is the relation of a note B to two other notes A and C which consists in the fact that any sound of A and any sound of C are separated by some sound of B: let us say that the note B *divides* the notes A and C.

The principle can now be stated as follows: *of any three notes, there is one which divides the other two*. This principle exhausts the essential content of the science of combinations of succession and resemblance in the world of sounds considered. All its other laws are deduced from this principle and from the properties of succession and resemblance in the same way as the theorems of mechanics and optics are deduced from the principles of these sciences and from the properties attributed to space and time.

Now these laws ascribe to the relation of division of notes the properties which define the *cut* relation or the *between* relation, the second form of linear order. The two forms are moreover equivalent and inseparable in the sense specified in Part One. In fact, from a relation with two terms (xy) which satisfies the axioms of the series, we can logically construct a relation with three terms (xyz) which satisfies the axioms of the cut: we need only set

$$\begin{aligned} A\,(xyz) \;=\; & (xy) \text{ and } (yz) \\ & \text{or } (zy) \text{ and } (yx) \end{aligned}$$

Conversely, starting with a cut relation (xyz), we can logically, but with slightly less facility, construct a series relation (xy). Therefore the two forms are indeed two definitions of the same order. This order of the simple open line, in its two inseparable aspects of series and cut is illustrated at first in the succession of individual sounds, and then in the division of classes of similar sounds or notes.

Let us, however, observe that the serial aspect is simplest for sounds, and that the cut, on the contrary, is simplest for notes. The succession of sounds, a simple relation, is in fact a serial relation: in order to obtain a cut relation, we must construct from it the relation of separation of two sounds by a third in accordance with formula A. On the other hand, the linear order of notes is most simply expressed by the relation of division, and this is a cut relation. Thus, of the two inseparable systems of linear *analysis situs* one of which is founded on a binary relation of directed order, the other on a ternary cut relation, the first is more simply applied to the series of sounds, the second to the series of notes. Here we already have an illustration of the observation in Part One concerning the relative character of extrinsic simplicity of geometric systems.

In this universe of auditive data, geometry – reduced, it is true, to the order of a simple open line – is illustrated twice: first, in the order of the sounds, next, in that of the notes. Nature very often embroiders the same pattern several times. She likes to illustrate the same type of order in the sensible texture in a manner first simple, and then complex: first in the sounds and their immediately apprehended succession, then in classes of sounds and the relation of division among three of these classes, a relation laboriously assembled by the mind.

These analogies, by means of which composites are symbolized by simples, are only with difficulty accepted as ultimate facts. How tempting it is to seek some common root for them in nature or the mind! However if our subject, busily philosophizing, thought he could see in the analogous orders of sounds and notes two expressions of the same fundamental fact, objective or subjective, would he not be mistaken? The order of sounds expresses a general property of sensible time, the order of notes a particular property of the path, which we would find easy to modify or destroy. If the notes, like the sounds, form a simple open line, it is purely accidental and does not indicate any unity. It is perhaps the same in our more complicated world; perhaps the multiple aspects of its geometric order are several distinct facts whose unity is only apparent in the light of invalid speculation.

The succession of simple external sensations of a conscious moving point which we displace along a line situated in an unchanging universe, therefore reflects *a few* of the geometric properties of the straight line, that is, its properties of linear order, through a linear order of classes of similar sensations. But the more particular property of zero curvature which determines the form of a straight line and distinguishes it from the parabola or zig-zag does not form part of this experience. In order to obtain a more precise idea of the type of rudimentary geometric properties which can be found there, and of the constitution of the terms and relations which illustrate them, we shall again consider briefly what would happen if the path ceased to be open and simple while still remaining a straight line (for suppressing this latter assumption would change nothing).

Any line. If we make our subject slide along a curve closed like an O or ramified like a Y, it is no longer true that, of any three segments, there is one which we must cross in proceeding from the second to the third. The principle of division of notes then vanishes; the previous system of laws governing the intermingling of successions and resemblances of sounds is destroyed.

But another more complex one takes its place. For there is a principle which enables the geometer methodically to decompose any line into a certain number of simple open segments, and which provides a formula for its constitution at the same time. This principle consists in *cutting off sections* of this line. If I cut off any section of a circle, a simple open line remains: this is the formula for the simple closed line. In the case of a ramified line, such as a Y or 8, it is necessary to cut the curve at its multiple points, or, in other words, for each one of these points we must cut off the section containing it. (We shall simplify the exposition by precluding the existence of singular points lying on the boundary of a section and at the same time of adjacent multiple sections.) The distinctive characteristic of all these sections consists in the fact that they are adjacent to more than two other sections: we shall call them *multiple sections*. Once these are cut off, the ordinary sections form a certain number of simple open branches. Each of these branches can comprise either one or two sections adjacent to a multiple section, and these two sections can moreover be adjacent to different multiple sections, or to the same one. In the first case, we have an isolated branch stemming from an intersection, such as the foot of the letter P; in the second, an interior branch, that is to say one which unites two intersections, such as the bar in the letter H; in the third, a loop departing from an intersection and later returning to it, such as the closed portion of the letter P. A complex line is characterized by the formula enumerating its branches and the multiple sections they touch. For instance, the formula for a Y or a T consists of three isolated branches touching a multiple section; that for a P or a 6 of an isolated branch and a loop touching a multiple section, that for an A or an R of two interior and two isolated branches each of which touches one of two multiple sections; and finally, that for an 8 of two loops separating a multiple section.

In place of a section, our subject cuts off a note, i.e. a class of similar sounds: he studies the *successions of sounds which do not contain any sound of a certain note*. If we assign him a path in the form of a simple closed curve, the law which prevails with respect to the blending of successions and resemblances is expressed in the following principle: *In the successions of sounds which do not contain a sound of a certain note, whatever this note may be, any three notes contain one which divides the two others.* Now let us make him describe a ramified path. He then distinguishes the notes with more than two neighbours (where two notes are said to be *neighbours* if their sounds directly follow each other): he calls them *multiple notes*. Referring to non-multiple notes as *ordinary notes*, he collects together all those notes whose sounds have succeeded each other once without being separated by the sound of any multiple note, so forming classes of ordinary notes which he calls *sequences*. *In each succession of sounds belonging to the notes of some given sequence, any three notes contains one which divides the two others.* Thus within the notes of the same sequence there reappears the order which before embraced the whole set of notes. Moreover each sequence contains either one or two notes neighbouring a multiple note which may be different for each one of the two, or the same. This results in three types of sequence: he calls them *isolated sequences*, *interior sequences*, and *closed sequences*. Now suppose that his path is a curve in the form of an H. We say that it consists of two multiple sections linked by an interior branch each of which gives rise to two isolated branches. Our character is no less learned, but he expresses himself differently and talks of other things. The universe, he says, is made of sounds; through their resemblances the sounds form classes which I call notes; and the set of notes consists of two multiple notes linked by an interior sequence each of which neighbours two isolated sequences.

An external experience marking a displacement along any path and containing only the relations of succession and global resemblance would thus provide a general principle of order and a form of the constitution of the world which faithfully reflects the general property of lines and the particular formula of the path in the way a book on *analysis situs* would state them.

Individuals, species and things. There is, however, an important difference between our subject's way of seeing things and ours. For us, the form of the path is an individual fact; but for him the formula which expresses it states a set of laws. For us, the *sections* of the line to which he is confined are indeed individual objects. But the *notes* which signify them are for him *species.* The relations among the notes are therefore relations among the species of his universe, and the classification of these into multiple notes and sequences should be compared, not to a geography, but to a systematic table of natural species like those constructed by the chemists; of course we must not forget the extreme simplicity of the species under consideration.

It is worth while to notice that this universe contains only individual sense terms and species, ephemeral *sounds* whose unity lies solely in their continuing existence, and which cannot make place for other sounds without being destroyed in the process, and *notes* whose members, on the contrary, are connected only by the relation of resemblance. Nothing yet answers to the type of entity we call a *physical thing.* Two sense terms separated in duration can in fact belong to the same physical thing; and, on the other hand, for this to be the case it is neither necessary nor sufficient that the two terms be similar. The relation by which we class them as 'appearances' of the same physical thing is really considerably complicated. This relation has not yet begun to be introduced into this initial schematism of a sensible world: we have taken care to reduce it to the simple relation of resemblance, which characterizes the logical type of the species, and we have done this precisely by excluding the difficult case of two 'things' which are indiscernible by the quality of the sense term which manifests them. Moreover things and species have been confused up to the present time.

Admission of indiscernibles. Let us try to advance a little way beyond the protection offered by the Leibnizian postulate. We shall forgo our constant identification of the different sections crossed with the different sounds heard. Let us, however, not permit two similar sounds to be associated with two sections which are so close that there is only one other between them: this is a *minimum of discernibility* below which all general structure vanishes in the world considered. But we shall allow certain sections separated by at least two others to emit identical sounds when we cross them.

We shall examine only the fundamental case of a simple open path.

If while walking along I see a telegraph pole and if, a little later, I again see a telegraph pole, the resemblance of the two sense terms is not enough to convince me that these terms are two 'appearances' of the same pole: for example, it would be necessary for me to have *retraced my steps* during the interval. Now what does it mean to retrace one's steps from a concise point of view which ignores kinesthetic sensations, as well as any change in the universe and many other complications? It means seeing the same landscapes in the reverse temporal order. This furnishes the complex relation which will be substituted for resemblance in the definition of the classes of sounds which indicate transits over the same section.

Briefly, this definition runs as follows. Let us call *simply symmetric* any succession of sounds distributed on either side of a central sound in such a way that each sound which precedes this central sound by a given number of sounds is similar to any sound which follows it by the same number of sounds. More generally, let us term *symmetric* any succession of sounds with the property that if we suppress in it all the sounds, apart from the first, of the simply symmetric successions it can contain, we are left with a simply symmetric succession. Similar sounds separated by symmetric successions indicate successive transits of the observer over the same section. Let us call such a class of sounds a *unity*.

The principle of order which has until now grouped sounds into notes now groups them into unities. This principle becomes: *of any three unities, there is one which divides the two others.* Linear order, therefore, continues to exist in a formally identical fashion on its own, but it is now associated with more complex terms. This will often happen to us: after grasping the order of an almost infinitely simplified universe, it will happen every time we inch one degree closer to reality by charging our sketch with a new feature, fatal to the previous order, but none the less an element of a more complex order of which the first appears to be nothing more than a particular case.

Let us again observe that, unless each note is common to more than one section of the path, certain notes will also be unities. This was the case for all notes under the hypothesis which excluded indiscernibles, and that is why the more complex concept of a unity did not emerge from the concept of a note. Since it was not required by any law, it had no opportunity to appear behind the simple concept of a *note* which concealed it so perfectly.

Now again, it is possible that the concomitance of the two notions remains the rule, and that there is only a single *note* which can be divided into several *unities*. But no matter: the whole import, and, so to speak, the whole weight, of the concept of a note, its whole importance as a nexus in the world, has passed irrevocably to the more complex notion of *unity*. The laws of experience which were previously associated with notes have degenerated into coarse rules which admit exceptions: the exact laws now apply to the concept of unity, and in this respect the note is nothing more than an inadequate approximation.

This concept of unity sketches the first outlines of what we call a *thing*. It enables us to see that the simplest form of a thing is incomparably more complex than the simplest form of sensible species. Moreover, it provides the first indication of the way in which this complication influences the sensible order observable by each one of us. As long as this order applies to the elementary sensible species, the uniting of two data in the same ordered class is brought about solely by virtue of the perceived resemblance between them, and there is no need to take into account what happened during the interval that separates them. On the other hand, when the ordering function passes from notes, which are pure species, to unities, which already have some analogy with things, the classification of two sounds in the same unity no longer results solely from their respective qualities, but also from the whole content of the sensible duration which separates them.

Surface or region. So far, the sensible universe we have studied expresses only the order of linear elements, not of a surface or of a space. But let us propel this observer along a surface divided into patches, or through a space divided into cells, each of which makes him experience a sensible quality which differs from those of its neighbours: the structure of the surface, or the space, is reflected in his experience. For on a surface or in a space as along a line there are, albeit with more alternatives, classes of elements (patches or cells) which cut all the paths connecting two given elements. Now properties of this kind are expressed, in the experience we are considering, by properties of the division of classes of similar impressions. Thus not only the order of a line, but also that of a surface and even spatial order can be expressed merely by the relations of succession and global resemblance of external data. Such an experience is, therefore, sufficient to illustrate the whole of *analysis situs.* On the other hand, the more particular geometry of straight line and distance has no place here: if the experience conveys the difference between a straight line and a circle, or a square and a cross, then it does not convey the difference between a straight line and a helix, or an angle and a curve. This shows incidentally that contrary to the suspicions of a certain philosophy, our method of analysis is perfectly innocuous. For it is quite powerless to introduce a complete geometry, as if by some sleight of hand, into any manifold whatsoever.

CHAPTER II

Succession and Global Resemblance

(kinesthetic data and data derived through any external sense)

The sensible universe we are about to study presents an ordered texture of the same general type as the preceding in so far as it is moulded from the *same relations*: succession and global resemblance. But it is of another variety inasmuch as its terms no longer comprise only external data but kinesthetic data in addition. The important thing here is that the difference between these two sorts of data does not in fact consist in a difference in quality, but in a difference between the physical causes which determine their similarities. For an observer wandering about an unchanging world, the resemblance between two external data signifies a return to the same place, while the resemblance between two kinesthetic data signifies a recurrence of the same change of place. The geometric order of the world explored will thus be expressed by diverse combinations of successions and resemblances, depending on whether these resemblances have external data or 'sensations of motion' as terms. Moreover, let us observe that our experience will involve the whole content of geometry books, and not merely the propositions of *analysis situs* as before.[(1)]

Consider a being endowed with a perfect kinesthetic sense moving unaided through a regular motionless medium where the effect of inertia is entirely absorbed, that is, whenever our subject's body interrupts the deformations which propel it, he stops in his tracks,

(1) This chapter may be regarded as a development of the ideas of Henri Poincarè concerning the role of sensations of motion in the experiential aspect of geometry. (*Science and Hypothesis*, Chap. IV). Poincarè claims that geometry is manifested in the alternation of external sensations and kinesthetic sensations: here we are trying to show exactly how. But he thinks that without kinesthetic impressions there can be no geometry in the sensible world: we have just seen, and we shall see later in more detail, that this exclusive dependence does not exist. In the spatial order of sensations of motion combined with external sensations, on the other hand, he assigns these latter a greater and more complex role than is useful: for instance, according to him the existence of changes in external impressions unaccompanied by changes in internal impressions is the essence of this order, whereas we shall see that is not the case.

like a mole which stops digging. This hypothesis avoids a considerable complication. Two similar deformations, i.e. two contractions starting from the same attitude and proceeding through the same attitudes with equal swiftness under these conditions result in similar displacements. For us, the life of this being consists of deformations interspersed with rests; for him, of the kinesthetic sensations thus produced. To be infinitely precise, two of these sensations have perfect global resemblance only when they signify either two identical deformations (and, consequently, two identical displacements) or two identical rests, i.e., of the same attitude and duration.

Moreover, let us assume that he apprehends each of his deformations and rests, no matter what their durations, only as a sensible whole, a single kinesthetic term, through the supposition that he does not discern more restricted terms therein. Thus, an arbitrarily long series of complex and varied contractions uninterrupted by stops appears to his consciousness as a unity that we shall leave undivided. He knows it to be similar to another sense term only when the latter reproduces it from beginning to end: between the sensations of two motions which have only one similar part (for instance, when one reproduces the other but is then extended beyond it), we take it that he will perceive a pure and simple difference only. Each of his motions, no matter how complex, can remain a total effort within which he discerns nothing. Likewise, we need not suppose that he perceives a general resemblance among the sensations of rest, which differ in attitude and duration, nor a general difference between these sensations and those which correspond to motions. There is a certain advantage in abstaining from assumptions of this sort.

Our subject has as yet only kinesthetic sensations. As long as the medium through which he travels is devoid of perceptible contrasts he will experience no other kind of sensation. He thus wanders about without ever meeting anything or getting anywhere. But so restricted an experience does not convey the geometric order of the world explored. The only observable laws are of far too rudimentary a type: they simply say that certain qualities of sensation never follow each other without an intermediary, i.e. those which signify a rest in a certain attitude or a motion terminating in this attitude, and a rest or motion starting from a different attitude. Thus a man walking cannot make two steps in succession with the same foot, because this movement takes his body from a certain attitude and leaves it in another. But incompatibilities of this sort undoubtedly do not make a geometry.

We are compelled to introduce external data to signify returns to the same places. But whereas in the preceding experiment it was necessary to assign indicative qualities to all the places traversed, here we can let all but one of them remain imperceptible. Let us, therefore, suppose that in one single place, attitude, and orientation, our subject experiences a certain quality of sensation, for example, a sound of a certain sort. The place, attitude, and orientation which determine the hearing of this sound will be collectively known as the *reference position.*

The laws of nature will consist in the recurrences of this reference sensation during the series of motions. All observations are thereby reduced to a single type: the assertion that such a sensation or series of sensations has separated two occurrences of the reference sensation, or else (in our language) corresponds to a motion or series of motions in a closed course. Is this enough to create a complex geometry? Granted that a sensible property distinguishes motions or successions of motions in a closed course from motions or successions of motions in an open course, does it follow that sensible properties constructed from the first distinguish, among the open course notions, those which effect the same displacement by different routes: among the various displacements, those which are translations; and among these latter, translations in the same direction or of the same length? This is what we are about to show.

But, you may ask, what is the use of all this economical refinement? After all, the real world gives us a perception, a sensible external reference, in each one of its regions, so why exhaust our ingenuity in showing that a single one of these references suffices? We do it in order to show that a plurality of sensible spaces are associated with this plurality of references, that these former are not necessarily in harmony, and hence that their plurality is a particular feature of the real world. Thus an analysis conducted with the greatest economy of means can extract, unaided, all the richness of the natural order.

Let us first of all restrict ourselves to the case of a plane geometry.

Plane motions. Let us compel the observer to move parallel to a certain plane – horizontally for example – and let us ascertain what laws he can discover.

We need a general term to designate the classes of similar sense data which signify either the same motion, or the same rest, or (for external data) the presence of a reference position. Let us call them *sensations*. A sensation will therefore be, not an individual datum, but the class of individual data similar to a certain datum (*definition* 1). The class of external data, all of which are similar to the subject, will be called the *reference sensation* (*definition* 2), and the individual members of this class will be called *reference sounds.*

For a certain motion to be possible immediately before and immediately after resting in the reference position, it is necessary for it to take the body from the attitude of this position and then return it to this attitude: the changes in place effected by such motions are therefore the same as those produced by freezing the subject's body in the reference attitude and then displacing it without deformation. These are the only motions we shall consider at first. In this experience they are signified by *sensations which can be preceded and followed by a reference sound.*

Let *rab*, . . ., *mnr'* be a succession of occurrences of the sensations AB, . . ., MN linking two reference sounds *r*, *r'*. The sequence AB, . . ., MN is said to be *closed* (*definition* 3).

If the interpolation of a sensation X between two terms of a closed sequence again results in a closed sequence, the sensation X is said to be *null* (*definition* 4). Such a sensation thus signifies either a rest, or a closed course movement equivalent to a rest.

In what follows we are going to consider the property of being closed which is possessed by certain sequences of sensations. It is really the only ascertainable property in this universe. The interpolation or omission of null sensations therefore changes nothing, and so we shall stop indicating them. Thus, the sequence XYZ, . . . will mean any sequence in which sensations X, Y, Z, follow each other in this order, either directly, or else separated by any null sensations.

Suppose X and Y are two sensations or sequences of sensations. If the sequence XY is closed, X and Y are said to be *inverse* (*definition* 5). In this case the motion corresponding to Y, executed after the movement corresponding to X, leads back to the point of departure.

Consider two motions B, C, both of which are inverse to the same movement A. They may differ in course and speed of operation, but they still produce the same displacement, i.e. that which nullifies the effect of motion A, and because of this they are equivalent. In an analogous fashion our subject calls *equivalent* two sensations or sequences of sensations B, C, both of which are inverse to the same sensation (*definition* 6).

We now see the experiential translation of the class of motions effecting the *same displacement*: it is the class of all kinesthetic data which are instances of one or other sensation *equivalent* to a certain one. This class, which is larger than one sensation, will be called a *unity* (*definition* 7).

Consider a motion A which, applied twice starting from the position of reference, cancels itself and leads back to this position. It is clearly equivalent to a semi-rotation of the body about a more or less distant vertical axis: indeed, this is the only horizontal displacement which is nullified by repetition. Our subject analogously calls a self-inverse unity an *alteration* (*definition* 8).

Now let us make the subject's body undergo two successive semi-rotations about vertical axes in such a way that the two resultant semicircles counterbalance each other, the result of which is plainly a *horizontal translation* (Fig. 5).

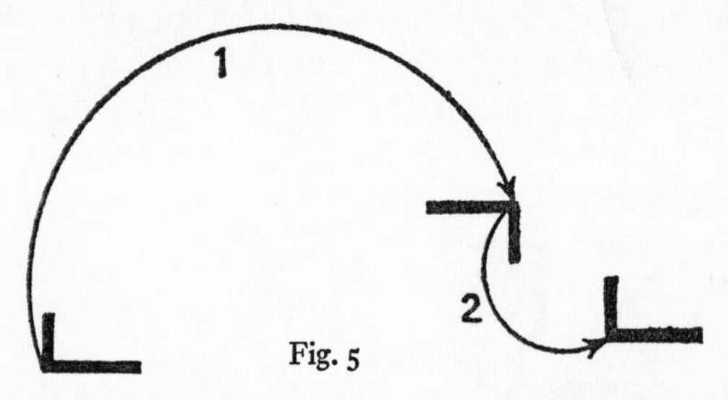

Fig. 5

Conversely, it is clear that we can transport the subject's body from any position whatever to any parallel position on the same level by means of two semi-rotations about vertical axes. Thus, the pairs of semi-rotations about vertical axes are equivalent to horizontal translations of the subject's body. He calls the unity formed by the sensations equivalent to the succession of two alterations *a double alternation* (*definition* 9). From what we have said, in his experience the set of double alternations reflects the set of horizontal translations of his body from the reference position.

Consider all the parallel positions this body can occupy as a consequence of these translations. These positions correspond to the points of the horizontal plane and present the same order as these latter. In particular, the congruence of two pairs of these positions (defined by the congruence of two pairs formed by any four similar points) obeys the same laws as the congruence of the points themselves. Since these positions are signified by the double alternations betokening the translations which separate them from the reference position, in order for the whole of geometry to be manifested in the sensible world considered it is therefore sufficient that the *congruences* among these positions be in turn reflected in some sensible relation among these double alternations.

Now this is precisely what happens.

Let us first of all see how the equality of two translations is reflected, meaning by this the equality of the distances through which they displace any point of the subject's body.

What relation among double alternations A, B which in experience reflects two translations A, B of equal length also expresses the equality of these lengths? The difficulty of establishing the equality of the movements A and B experientially derives from the divergence of their directions. However, this problem can be overcome.

Consider any translation A and a motion M which is not a translation and changes the observer's orientation by an angle φ (Fig. 6).

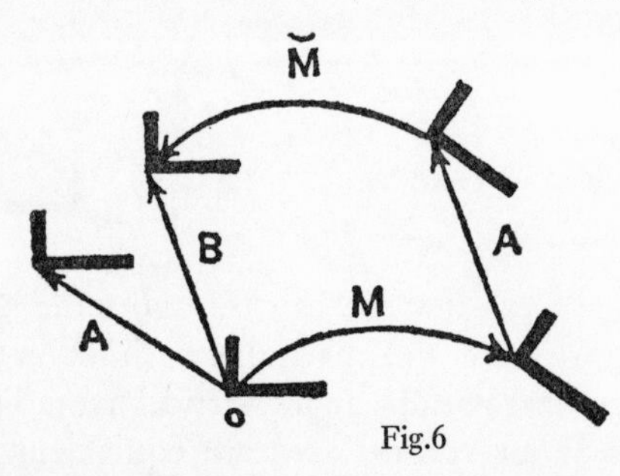

Fig.6

Starting from the reference position o, let us execute motion M, then translation A, and finally a motion M̆ inverse to M: the net result is clearly equivalent to a translation B *of the same length as* A *and at an angle φ with* A. Conversely, any translation B which is equivalent to a sequence consisting of a motion M, a translation A, and the inverse of M has the same length as A. Finally, for any two translations A, B of the same length, there exists a motion M such that the sequence M- A- *inverse* of M is equivalent to B.

This furnishes a definition of equality of two translations which is expressed in terms of the relations between the corresponding double alternations: our character defines two double alternations a, b to be *equal* if, for some unity m, the sequence mam̆ is equivalent to B, where m̆ is the unity inverse to m (*definition* 10).

But what we seek is an expression for the relation between two pairs of translations AB, CD whose positions of arrival are separated by the same distance *d*. Such an expression can easily be obtained. Let us define the *difference* of two translations to be that translation which proceeds from the point of arrival of the first to the point of arrival of the second: it is clear that the differences from A to B and from C to D have the same length. Our subject who has analogously assigned the name *difference of the two unities* A, B to the unity X with the property that the sequence AX is equivalent to B (*definition* 11), finally arrives at the following definition: when the differences of the double alternations AB and CD are equal, I shall say that the two pairs AB, CD are *connected*.

Our object is thus achieved. The connected pairs of double alternations AB, CD in fact correspond to the pairs AB, CD of translations such that the pairs *ab*, *cd* of positions of arrival of these translations from the reference position are congruent, and we know that the laws governing the congruences of parallel positions are the same as those governing the congruences of points. Plane geometry is thus wholly reflected in our subject's sensible universe. If he knows all that can be experientially known about this universe, and if he then recalls a *Plane Geometry* which we suppose him to have read, he notices that by taking *point* in the sense of *double alternation* (*definition* 9) and *congruence* of two pairs of points in the sense of *connection of two pairs of double alternations* (*definition* 12), all that the geometer says holds true for his own sensible process.

Movements of any kind. We now pass on to the case of motions which are no longer necessarily confined to a horizontal plane, but of any kind at all; let us see whether sense data continue to present a geometric structure under these conditions.

Some of the notions we have considered are not affected by passing from plane motions to movements of any kind at all. Examples are: the *sensations* (*definition* 1), the class of *reference sounds* (*definition* 2), the *closed* sequences (*definition* 3), *equivalent* sensations or sequences of sensations (*definition* 6), the *unity* (*definition* 7), all of which invariably signify the same physical entities.

Likewise, *alternation* (*definition* 8) retains its meaning. For, of all the motions a body can execute, the semi-rotations about an axis are the only ones which, when repeated, return the body to its original position. However, the axes of two semi-rotations are no longer necessarily parallel as they were for motions in a horizontal plane. Consequently, the net result of two successive semi-rotations is no longer necessarily a translation since the two half-turns do not counterbalance each other as they did before: *double alternation* (*definition* 9) no longer furnishes the sensible realization of translatory motion. We must try to discover such a realization, if indeed there always is one.

Consider a semi-rotation A. For what semi-rotations X is it true that A followed by X produces the same displacement as X followed by A? We state that a necessary and sufficient condition for this equivalence of XA and AX is that *the axis of* X *cut the axis of A perpendicularly.*(1)

(1) In fact, let *o* be the reference position, *a* the position of arrival of A, *x* that of X, *ax* that of AX (and, by hypothesis, that of XA) (Fig. 7).

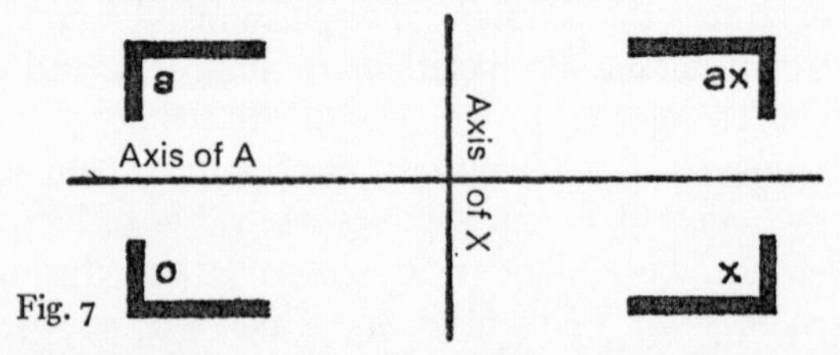

Fig. 7

The semi-rotation from *x* to *ax* is the semi-rotation X from *o* to *x*, but it is performed after its axis has been shifted by the semi-rotation A. Therefore, *x* and *ax* are situated symmetrically on either side of the original position of the axis of A. But in order to get from *x* to *ax* we must perform semi-rotation A after its axis has been shifted by the semi-rotation X. For this to be the case it is necessary that *x* and *ax* be situated symmetrically on either side of the position assumed by the axis of A under the influence of X. Now two volumes cannot be symmetrically situated with respect to more than one line. Hence the positions of the axis of A before and after X must coincide, and this occurs only when the axis of X cuts the axis of A perpendicularly.

Therefore, if A and B are two non-equivalent semi-rotations, that is, about different axes, all the semi-rotations such that AX and XA, BX and XB (if they exist) are equivalent have axes perpendicular to the axes of both A and B; they are, consequently, all parallel, and a sequence of two of them MN is equivalent to a translation.

Our subject then formulates the following notions. Two alternations X, Y are *permutable* if the double alternation XY is equivalent to the double alternation YX (*definition* 13). If there exist two alternations M, N both permutable with the alternations A and B, the double alternation AB is said to be *homogeneous* (*definition* 14).

Just as the positions obtained before by horizontal translations of the observer's body from the reference position were expressed in his experience by double alternations, the positions now obtained by *any translations whatever* of his body from the reference position correspond to the homogeneous double alternations which have just been defined. As for the congruence of two pairs of these positions, it is expressed, as before, by *connection* (*definition* 12) of two pairs of double alternations, through the notions of *equality* (*definition* 10) and *difference* (*definition* 11), of two double alternations, provided we read *homogeneous* double alternations everywhere.

Geometry is thus wholly expressed in this experience. *Homogeneous double alternations* and the *connection* of their pairs provide our subject with an interpretation of *points* and *congruence* in every statement geometry makes about them; and we know that a full formulation of geometry can be given in terms of these expressions alone.

We observe, however, that the axioms which form the simplest basis for geometry when it takes *points* and *congruence of points* as primitive terms lose this privilege as soon as we ascribe the complex significations *homogeneous double alternation* and *connection* of *two pairs of homogeneous double alternations* to the purely formal terms *point* and *congruence*. In this case the simplest axiom assumes a very complex meaning. Indeed, *two pairs of points congruent to the same pair are congruent to each other* now means *two pairs of homogeneous double alternations connected with the same pair are connected with each other*. Now any verification of this formula in the experience under consideration would involve some fifty sense data at least. But on the other hand this statement can now be inferred from simpler statements, such as the one which asserts the symmetry of inversions: *If the sensation* A *is inverse to the sensation* B, *then the latter is in turn inverse to the sensation* A. And these are the axioms of the new system. It would be interesting to investigate them in order to see the particular formal appearance the structure of geometry assumes in these circumstances. But this would be a considerable task, and we shall not undertake it here.

Other spaces. Does the logician whose experience we are studying find therein only *one* illustration of geometry, only *one* possible interpretation of points and congruence? Far from it: he discovers a large number of solutions all of which are independent of each other, as indeed they are of the solution we have just witnessed, which is merely the simplest of all.

It must be remembered that among the subject's motions we have considered only those which start and finish in the attitude of the reference position. The expression of geometry we have just extracted was based entirely on the kinesthetic data which reflect these particular motions.

Now let *p* be a certain position of the movable body which differs from the reference position *o* in both place and attitude. It is clear that the motions which start and finish in *p*'s attitude would be conveyed in the subject's experience along with the geometric order we have just seen if *p* were a second reference position signalized by some recognizable impression. It is in fact nothing of the sort, but the result is the same nonetheless. For if there is no sensible indication of the subject's presence in position *p*, there exist determinate motions which proceed from this position to reference position *o*, as well as other motions, inverse to the preceding ones, which lead back from *o* to *p*. Let D, D̆ be a mutually inverse pair of these motions, and let D, D̆ be the kinesthetic unities which translate them into experience. As 'references' let us take the sequence DRD̆ in place of the class R of reference sounds. A kinesthetic datum *x* which follows one of these new references expresses a motion which has started, not, as before, from position *o*, but from position *p*. Similarly, a datum *y* which precedes an occurrence of the reference sequence DRD̆ expresses a motion which has terminated in position *p*.

Thus a new formulation of geometry makes its appearance. Its motions will be defined as before. Apart from the replacing of references R by complex references DRĎ, homogeneous double alternation and similarity remain the same. They illustrate afresh the axioms of the geometry of point and congruence. But this illustration, although similar in form to the preceding, is none the less independent. Its content is entirely new. Not one of the kinesthetic data it puts into play enters into the preceding system, for the new system's material consists of the data which reflect motions starting and finishing[1] in the attitude of position *p* (which differs from the reference attitude), whereas the first interpretations were based exclusively on the kinesthetic translations of motions starting and finishing in the reference attitude.

For our subject, therefore, the geometry based on references DRĎ is not deducible from the geometry based on references R. The fact that the same formal laws continue to exist after we have replaced R by DRĎ, thus completely renewing the sensible content of the notions of homogeneous double alternation and similarity, is for him only a *de facto* harmony, which rightly surprises him.

But any other pair of mutually inverse unities MM̌ yields the same result as DĎ: the adoption of the sequence MRM̌ as a reference organizes geometrically an entirely new mass of sense data, which are completely foreign to the two preceding 'geometries' (provided they do not include M itself). The independence of all these interpretations is absolute, and their quantity inexhaustible, since they are as numerous as the attitudes the subject's body can assume.

A set of terms ordered by a binary relation according to the euclidean axioms of congruence of pairs of points will be called a *space*. We must admit that the subject knows a multitude of similar but distinct spaces formed by different sets of sensations. We have just seen that their structures are independent and hence form so many primary facts. However, they do manifest a certain connection which is itself another fact.

Formation of a total space. Let E and E′ be two of these spaces. Their points are homogeneous double alternations, i.e. particular

(1) Except the data D whose corresponding notions start from the attitude of the reference position and data D whose corresponding motions terminate therein. But these motions are also foreign to the first system, since they permit, either at start or finish, a different attitude.

classes of kinesthetic data. In general these two spaces do not have in common any of these classes nor even any of the data which comprise them. They are two sensible wholes exterior to one another. We know that they reflect two sets of motions which differ in initial and final attitude. Let A be the motion leading from the position of origin of space E to the reference position; let A be the corresponding unity and Ă its inverse. The reference of space E is formed by the sequence ARĂ, where R denotes the class of reference sounds. Similarly, if A′ is the motion leading from the origin of space E′ to the reference position, A′ the corresponding unity and Ă′ its inverse, then the sequence ARĂ′ is the reference of space E′. The motions reflected by the kinesthetic space E are the translations effected by motion A starting from the reference position. Let T be any motion in space E, T′ any motion of space E′, and M a motion leading from the position of arrival of T to that of T′. This motion M enjoys the following property: performed after *any* motion X in the space E, it terminates in the position of arrival of a certain motion X′ in space E′. Moreover, the resulting correspondence between motions X, X′ of the two spaces E, E′ preserves the relations of congruence: if, as a result of applying M, the motions X_1, X_2, X_3, X_4 correspond to X'_1, X'_2, X'_3, X'_4, and if the two pairs X_1X_2, X_3X_4 terminate in congruent pairs of positions, then so do the pairs $X'_1X'_2, X'_3X'_4$.

This property is experientially reflected by the following laws, in which the small letters denote the 'unities' which kinesthetically express the motions denoted by the corresponding large letters, and the unaccented and accented letters denote homogeneous double alternations or 'points' of the spaces E, E′ whose origins are separated from the reference position by the motions A, A′ respectively. If

AXM is equivalent to A′X′,

then *any* sequence AYM is equivalent to a sequence A′Y′ and conversely; furthermore if the pairs ST, UV of the space E are congruent, then the same is true of the pairs S′T′, U′V′ of the space E′ which correspond to them through the unity M.

Thus, an application of the same unity M to the points of E transforms them into points of E′ by preserving the congruences, and consequently all the geometric properties, of the sets of points. We can also say that the space E′ is applied to the space E via M. Similarly,

E″ is applied to the space E′ via the unity M′; E‴ is applied to the space E″ via the unity M″, etc. All these kinesthetic spaces are applied to each other in such a way that their ordered textures coincide. Thus let us call the set of all points which are applied to one another a (new) *point*, and the relation between two pairs of these new points, whose members form two congruent pairs of old points in each space, the (new) *congruence* relation. This new space incorporates all the preceding ones; it fuses them into a single structure. Through the connections he discovers among all the kinesthetic spaces, our subject sees them superposed to form one total space.

But this composition is quite arbitrary: this total space can be constructed in a multitude of incompatible ways, and not one of them is more prominent than the others. Indeed, consider two of the spaces it unites, E and E′, say. Let T be a point of E, and let τ be the point of the total space to which T belongs. Which point of E′ belongs to that same total point τ? Answer: the point T′ into which T is transformed by a certain unity M. Now this unity can be chosen in such a way as to 'apply' the given point T of the space E to *any point* T′ *of* E′ *whatsoever*. Thus, the various kinesthetic spaces E, E′, E″ are genuinely applied to one another, forming a total space if you wish; but nothing tells us what points of these spaces together form a given point of the total space. The total space of motions, therefore, remains an indeterminate construct in this situation: although far from perfect, this is the best description in this universe of the space we regard as embracing all nature in a single geometric structure.

A *Geometry* would indeed be an astonishing book for our subject. The constant recurrence of the unknown terms *point* and *congruence* (assuming only these) makes it an enigma in itself, and on top of this it admits, not just *one* sensible interpretation – which would already be a remarkable fact – but a whole throng of independent interpretations. In addition to the reference sounds, all of which are similar, the sensible universe contains an infinite variety of kinesthetic data. In the midst of the chaos appear first one mass, then a second, then yet others which illustrate geometry, each one composed of data which reflect the motions taking the body from and leaving it in a certain attitude. The laws governing these different masses are therefore of the same form, although the systems themselves differ in both the 'references' they posit and in the diversity of the very data they involve. They assign as many different significations – all of them true – to the set of axioms and theorems. Each furnishes a complete and independent interpretation of the points and their congruences. Each forms what we have called a *space*, and all these spaces are in turn linked together by uniform transformations.

If we were in the subject's position, knowing what we know, or what we think we know, we would undoubtedly think we could *explain* this indefinite recurrence of the same type of order in nature, and reduce it to a single fact. But we are restricting ourselves in this study to *analysis*. It is, therefore, advisable not to go beyond the complexity of the geometric textures it extracts, for even if this complexity did spring from a simple source which could be seen by the philosopher, it would nonetheless remain intact in the universe of experience.

CHAPTER III

Introduction of Local Diversity of Sense Data

We have just studied the first two types of geometry of sense data. Both have the same relations as elementary links, namely *succession* and *global resemblance*. But they interlace differently since the resemblance of two sense terms in one case signifies the observer's presence in the same place, and in the other, the same propulsion of his organs. In an experience restricted to these two relations alone, the first example has exhibited an illustration of geometry in the data received through any external sense, and the second, in the data of the internal sense of movement and attitude.

But the first of these geometries, the external type, is not complete; and the second, on the other hand, is not a purely internal type. The succession of external data (sounds) we studied at first expressed only that amorphous part of geometry called *analysis situs*. On the other hand, if the mass of kinesthetic data later examined has formed a complete geometry, it could only have done so with the aid of a minimum of external references. We might be tempted to conclude from all this that the co-operation of several senses, and, more particularly the coalition of external and internal data, is essential to the expression of geometry in experience: this is the opinion reached by Henri Poincaré. But we shall see that this conclusion was drawn too hastily.

Did the imagination not resist the mind's attempt to make these first two sensible geometries operate within a nature without any other content? The melody we first posited, the flux of kinesthetic data we later considered, strike the mind as an excessively narrow, ephemeral reality, spirited away by a succession which is far too simple to lend itself easily to reflection and analysis. It is disconcerted by the absence of a certain structure so familiar to us that it has become bound up with our very idea of a sensible universe.

Of the two fundamental ordered networks in our experience, only one was in fact present, namely that which the pure and simple recurrences of things form through the connections of global resemblance running across succession. But there is another of which we have not yet spoken. This is the network formed for a sense such as vision by the two *diverse resemblances* of colour and immediate position. Global resemblance is divided into qualitative and local resemblance. Whereas in the preceding universes an auditive or kinesthetic term belongs to *only one* class of terms arranged by resemblance – a class which we have sometimes called a *note*, sometimes a *sensation* – a visual term belongs to *two* classes of similar terms simultaneously, the one being the class of terms which resemble it locally – let us call this first class its *sensible locality* or *position* – the other being the class of terms which resemble it qualitatively – let us call this second class its *quality* (in the narrow sense).

The network of resemblance and succession is now complicated by intersections of these two types of resemblance. A sound is nothing more than a recurrence of its 'note', a kinesthetic datum, of its 'sensation'. But a visual term is a coalition of its 'locality' and its 'quality'. A note either is or is not present at a given instant, but a colour either is or is not present *here* or *there*, in such and such a place in the visual field. The flux of sensation, instead of having as its structural elements pure and simple recurrences alone, becomes a game in the course of which two kinds of entities, namely the sensible qualities and positions, coalesce and separate in a thousand ways.

Moreover, although the qualities occur irregularly, allowing themselves to be forgotten for long periods, any sensible position, in the visual field for instance, is continually present to me through some datum: a stone, a leaf of a tree, a shadowy region, a section of sky, and so on, from morning to night. Qualities – red, blue, green – come and go, but the sensible positions are always manifested in their entirety. This gives them a certain privilege. As an ever present whole they form a kind of background which throws the qualities into relief, a canvas over which they range, a stage on which they perform as a company. A visual term does not vanish completely like a sound: it leaves behind it an heir to one of its characteristics. Its sensible locality survives it in a new term. This detracts somewhat from the absoluteness of the flux. The appearance or disappearance of a datum only succeeds in realizing or cancelling a certain possible arrangement, one of whose two factors, the sensible position, the *here* or *there*, always remains present to me.

This play of two diverse resemblances across vast sets of simultaneous terms, this combination of two groups of characteristics, one of which is always represented in its entirety, is what characterizes the visual universe, and it is this we miss so strongly in the more rudimentary senses. This is what makes them appear so meagre, so contrary to thought. Our imagination cannot stop there. As for the physical displacements which have provided me with external definitions of the sequences of sense terms I wanted to consider, have I not represented them to myself in a quasi-visual form? Have I not imagined seeing the subject's body first here, then there, in some field similar to the visual – in other words, have I not imagined myself apprehending a certain sensible quality changing its position?

Besides, one sometimes sees the apprehension by the mind of a field within which it can arrange various terms in various places as the indispensable instrument for all intellectual construction, the ideal blackboard which alone enables it to contemplate an order. It would then be necessary to furnish our preceding subjects with imaginations essentially different from their auditive and kinesthetic experiences, or at least infinitely more subtle, since that experience admits no partition of resemblance into local and qualitative. We will not be permitted to imagine the orders of succession which we have only analyzed by contemplating them in a simultaneous vision amid an imagination similar to sight for the diversity of *heres* and *theres* it presents, and whose intuitive structure would be the fundamental manifestation of geometry.

But first of all let us see what this new type of sensible structure, in which succession no longer plays a role, really is.

CHAPTER IV

Relations of Position

(*visual data*)

The succession and recurrence of sensible global qualities within the narrow limits of an experience in which the data follow each other in a single file already form two types of geometry, as we have seen. However, we do not naturally imagine the geometric structure of nature in this way. To the imagination a space is not a *set* dispersed in time, but rather a simultaneous multiplicity. Indeed, if we consider the geometric order of the all-changing world of physical realities, it can only be understood through its order at a given instant. Moreover, everyone admits this: however, the relativists think that simultaneity and, as a result, the geometric order of physical things varies with the system of reference. In physics, spatial order is in its very essence an order of simultaneity.

But the simultaneity involved is not the immediate sensible relation I call by the same name. Let us in fact imagine a large number of events occurring, from the standpoint of physics, at the same instant in all regions of the world. Let us say that these *physical events* are simultaneous: but I cannot perceive them simultaneously. Of the *sensible events* which manifest them, some occur earlier, others later, in my sensible time. What the physicist calls the order of space is inseparable from what he calls simultaneity – but what he calls the latter is not sensible simultaneity.

However, I have the (possibly chimerical) notion of a space whose terms, immediately simultaneous for me, would present an intuitive order free of all succession. The family of *relations of position* of sense data furnishes some substance for this dream: for if these relations really do have simple natures, then they, in a degree great or small, realize the idea of intuitive and instantaneous geometric order.

In a single glance I scan the star-studded sky. Among the triads of stars, I can discern those whose terms appear to be in alignment. I can classify the triangles formed by three unaligned stars into scalene, isosceles, etc. I can see that these two angular distances are equal, and that these two others are not. Of course, these classifications would be more reliable if I were to use instruments. But even with the naked eye I can roughly perceive the resemblances and differences among the figures formed by the stars in various regions of the sky. Are we not dealing with some kind of instantaneously generated geometry here?

Let us admit, provisionally, that these relations of position among the sparkling points whose dispersion throughout the sky is apprehended in a single glance are in fact relations entirely enclosed in one instant, without intrinsic reference to past or future perceptions. By doing this we are assuming the existence of a sensible intuitive geometry. Let us consider the extreme case in which this geometry is as simple, as complete, and as perfect as possible.

Imagine each point of physical space occupied for the time being by a material point, and a motionless observer who can take in all these points distinctly at a glance. We are thus conceiving an extremely idealized 'vision' which in a single act surveys, without movement, the whole content of the universe. No object conceals another object from him; and every point is discerned. The material points which occupy the space filled by the observer's body are no exception: like the others, they are distinctly 'seen'. As for the body itself, we may imagine it to be made of an imperceptible and penetrable matter which is non-existent so far as the vision is concerned. The mind whose existence we are assuming is thus confronted with an infinity of distinct and simultaneous sense data, corresponding term for term to the points of our physical space. This is the greatest possible intuitive *perception* of space.

Moreover, he immediately apprehends the relations of position among these data. Let us simply consider congruence: among all pairs of sense terms aa', bb' corresponding to pairs of luminous points $\alpha\alpha'$, $\beta\beta'$ which themselves occupy two congruent pairs of points in space AA', BB', and among these pairs only, he perceives, in a single glance, a certain indefinable resemblance we shall call connection. This is the greatest possible intuitive *knowledge* of space since all geometry can be stated without employing any relation other than congruence. Therefore, we cannot imagine a more perfect immediate knowledge of the spatial order of this sensible world.

Any geometric statement is thus expressed by a sensible law. Points become the *minima visibilia* (for we are pre-supposing infinitely fine vision), and congruence becomes the intuitive connection among pairs of them. For instance, the axiom which says that congruence is transitive now says that connection is transitive, and so on. For this subject, as for its predecessors, geometry expresses the laws of nature: not more truly, but more simply. For on the one hand, it is manifested in his experience only in a single order on a single occasion, and not in diverse orders astounding in abundance. On the other hand, the fundamental geometric relations – let us say, for simplicity, the congruence of pairs of points – previously admitted extremely complex significations only: recall the connection of pairs of homogenous double alternations. In the present case, congruence admits as a 'value' a simple line, a relation that is basic and no longer composite. Finally, the sense terms it connects, instead of being scattered in duration, are present simultaneously, together forming a single vision. This provides a full explanation of why this type of expression of geometry in experience seems the most natural. Thought refers to it so involuntarily that we may well have to ask ourselves whether such a display of simultaneous terms in a field does not constitute, for our image-bound minds, the basis for all geometric order, or even for any order whatsoever.

Let us pause for a moment to see how this point of view influences the analysis we are pursuing. It means interrupting the exposition of the analysis. But must we not attempt to defend its very principle against a doubt which is prevalent today?

CHAPTER V

The Significance of the Hypothesis of a Natural Spatial Symbolism

We have endowed our subject with perfect intelligence without specifying any conditions for its exercise. Now we must admit that it functions only with the aid of an instinctive or deliberate representation of its objects and their relations on some ideal blackboard.

We can conceive this auxiliary medium as the image of a tactile or visual field, these two senses furnishing a *here* or *there*, a clear empirical distinction between quality and immediate position, or else as the image of any field of sensation, thus investing the mind with the power of distinguishing quality and place, even when they are invariably bound together, and with a capacity beyond all experience of representing a qualitative term as situated *here* or *there* in the auditive, olfactory, or kinesthetic fields. Finally, we might conceive the domain of *heres* and *theres* as forming a background for the intelligence, like a kind of innate image, a proper object of an *a priori* intuition. All these ways of looking at the matter amount to the same thing for what follows. But in order to clarify our ideas, we shall speak as if we were dealing with a visual schematism.

Suppose then that the analysis of an auditive or kinesthetic reality, such as those we have just sketched by ascribing them to our subjects, cannot be performed directly, but only through the projection of this reality into the midst of a visual imagination which presents an intuitive order of immediate localities. Does it follow from this that the analysis which has on two occasions exposed a manifestation of geometry in the succession of data occurring one at a time, really applies only to the field used as a background? Would it have extracted only the laws of this obligatory medium of representation?

We seem to have just deprived ourselves of all means of proving the contrary. However, this is not the case.

The concept whose significance we are studying rests entirely on the notion of *representation* or *symbolism*, and it is through this that we must examine it.

If imagining a fact by means of a symbol consisted in thinking 'here is the fact in question' in the presence of a certain imaginary background, if the relation between the symbolized thing, which is by hypothesis inconceivable in a pure state, and the symbol, which alone can be imagined, effected a complete substitution of the symbol in all its aspects for the thing, then the examination of things through the aid of symbols would in fact be nothing more than the examination of the symbols themselves.

But this is hardly the case. All symbolism is partial or at best abstract. In any symbol, the aspect which symbolizes something is accompanied by other aspects which symbolize nothing which the mind ignores. In the letters of the alphabet, the form alone is symbolic of the sound: the colour of the ink, the dimensions of the strokes signify nothing. The order of words inside a line represents an analogous order of sounds. On the other hand, the vertical order of words is a formative accident and symbolizes nothing. Now in China the opposite is true. This illustrates the fact that from the same images arranged in the same way the mind *selects* the qualities and relations to which it assigns the import of a symbol. In the presence of an unknown script, one cannot begin to investigate the signification of the symbols which appear in it without having made some conjecture about what constitutes a sign. For we do not know, and it is important to know, whether for instance the colour or thickness of the strokes are accidental or essential to the symbol.

Of course we are sometimes tempted to *extend* a symbolism. We inquire whether, by some chance, new aspects of the images employed as symbols might not also be of service. For example, after representing the chemical composition of bodies by collections of atoms, we have subsequently made the relative positions of the atoms enter into the symbolism. This kind of development occurs frequently in the history of the sciences. It is a research procedure. But then to lose sight of the fact that every symbol is imbued with irrelevant aspects in order to postulate that everything in it signifies something, is to retreat from reason into folly. This mystic state of mind is perhaps the origin of thought through symbols, and perhaps also its failing.

But it is not the common method, except perhaps among some savages.

A symbol is, therefore, only a symbol in respect of those properties distinguished by the mind. And likewise, the thing symbolized is not symbolized completely, but only in certain of its properties. True, according to the hypothesis of this discussion, I can only think about a thing through the aid of its symbol. But, without abandoning the condition thus stated, I claim that I can see that a property A of a symbol does not represent any aspect of its object to me, and that, on the contrary, properties B and C do represent something to me. Furthermore, I can see that these two signifying properties *do not* represent *the same thing to me*. For instance the form of the letters in an italicized word indicates the general sound of the word, and the type face indicates a more accentuated intonation. The shade of the paper indicates nothing; the pitch or rapidity of the sound is not indicated by anything. Thus each of the symbol's symbolic aspects is incontestably associated mentally with a determinate aspect of the thing symbolized.

Just as there is an element of chance in the meaningless part of a symbol, so there is an element of necessity in the appropriateness of its symbolic properties to the properties they symbolize. I have just heard four o'clock strike; on this day the sound of bells is projected into my visual schematism in the form of four collinear points. I perceive that the number of these points can represent a certain aspect of what I have heard, which I shall call its plurality. Their order from left to right can represent another aspect of the sound of the bells, which I call its succession, provided that the symbols of the different strokes are arranged in a certain way and not in another. Finally, if the strokes were all of different notes, the order of points along the straight line represents the order of pitch, only when the pitches of successive strokes ascend or descend continuously. In other cases, we need two different alignments to symbolize the order in time and the order of pitch. Or else, if the brilliance of a point happens to be correlated with the sharpness of the note it represents, I establish that the order of brilliance can represent the order of pitch. I observe it, but can do nothing about it. All this concerns a possibility or impossibility which is independent of my mental decrees.

Thus my symbolic representations, even if I cannot dispense with them, instruct me without blinding me, for they allow recognition of symbolic properties and perception of the appropriateness of these properties to certain aspects of the thing represented.

What then is this appropriateness? What relationship between two properties makes one a possible symbol of the other? Or else, to remain within this same area of symbolism, *what relation is there among all possible symbols of the same real property?* I can represent the plurality of the sounds by an infinity of diverse visual schematisms. What do they have in common? Nothing apart from the number of their terms. I can represent the general relation of sensible succession by the order of a straight line, a parabola, a sinusoid, a helix, and many other lines, arranged on my mental blackboard in any manner whatsoever. What do they have in common? Nothing apart from the fact that they are all open simple lines. We can generalize these examples as follows. Among all the relations capable of symbolizing accurately the same aspect of the reality studied, there exists an abstract relation of *formal analogy*. In order for the relation R′ to symbolize the same thing as the relation R, we shall in all cases find it necessary and sufficient that these two relations be *similar*, i.e. they have the same formal properties, or again that it is possible to replace one by the other and the field of one by the field of the other, *salva veritate*, in any proposition containing in addition only logical or mathematical expressions.[1]

When I recognize that a certain aspect of the reality I am contemplating can be accurately represented by the relation R, it is therefore the formal type of R whose appropriateness I recognize. Thus only the asymmetric, transitive, and connected type is fit to represent immediate succession. But the properties of nature which we must consider in our sensible geometries are themselves formal properties. Hence the introduction of any schematism, visual for example, only produces the following change. Instead of thinking: 'In my experience there is a relation which I call succession and which *is* an asymmetric, transitive, and connected relation', we shall think: 'In my experience there is a relation which I call succession and which *can be accurately represented by* an asymmetric, transitive, and connected relation', and similarly for the rest.

(1) Cf. L. Wittgenstein, *Tractatus logico-philosophicus* (London, 1923).

Analysis of the notion of symbolic representation thus limits the philosopher's employment of it in his critique of the truth of ideas. This limitation is marked by three facts. First, it is not the concrete image which symbolizes, but one or other of its properties. Secondly, it is not the concrete thing that is symbolized, but one or other of its properties. Thirdly, the appropriateness or inappropriateness of an intuitive relation for symbolizing something, a determinate something, is an apprehensible fact associated only with the formal type of this relation. These three characteristics constitute the abstract nature of all symbolism, thanks to which the mind by the use of the symbols really grasps the *thing*, and not simply the symbol in place of the thing.

Even if our subjects had been able to contemplate the relations among their kinesthetic or auditive data only by projecting them into a visual imagination where the geometry is expressed by an intuitive order of position, the geometric order they have discovered in these data would not have been the effect of this projection. For the geometric order could not have symbolized anything when surrounded by the properties of this symbolic medium. In part, this is what actually happens in the first example. If the notes signifying the observer's displacement along a straight line were projected for him into the sections of a visual line, it would be a pure accident. For only the open and simple order of this image would represent a fact of nature to him. Its rectilinearity would represent nothing to him, and the difference in aspect between a straight and wavy stroke on his ideal blackboard, although familiar to his imagination, will *represent* to him no feature of his experience whatsoever. Likewise, if his path is ramified in the shape of a Y, he will be able to picture it indiscriminately as a Y, a T, an E, an F, since the order common to all these forms and many others is to him only symbolic of the fact which he can grasp.

In this study of sensible nature we therefore need not be preoccupied with any hypothesis about the role of a schematism presenting an intuitive geometric order. The foregoing analyses are not affected by it; for the schematism, as it actually appears, is merely an instrument of the reason. It remains the servant and not the master. It in no way determines the sensible facts that reason perceives through its aid and which are our sole concern here.

CHAPTER VI

Relations of Position, Simultaneity, Qualitative Resemblance, Local Resemblance

(visual data)

We have come across two kinds of sensible geometry: the geometry of succession and global resemblance, and the geometry of relations of position. There exists yet a third, namely the geometry of simultaneity, local resemblance, and qualitative resemblance. It is the most interesting because it best recalls the methods of the physicists. However, let us approach it indirectly. Let us first see what nature would be like if the network of simultaneities, local resemblances and qualitative resemblances were not isolated, but conjoined with the network of relations of position which has just been introduced. This in fact realizes the maximum of an intuitive geometric order, the very ideal of an immediate apprehension and immediate science of space. It is worth our while to notice all the former adds to it, however, and later on we shall show that this first network suffices.

Imagine, then, the same physical world and the same spectator as before. But instead of allowing this spectator to remain motionless, as we have done up to now, let us successively place him in different positions. Suppose, for greater simplicity, that his perception is completely obliterated during the displacements, like that of a man who may only be moved around passively with his eyes blindfolded. The single scene which formed his complete universe thus gives place to a second, then to a third – to a series of ephemeral scenes which succeed and replace each other.

The experience becomes open and expanding. The world splits into fragments and multiplies: what new order does it contain?

Reality is presented in it by large aggregates which succeed each other like so many transitory worlds. Let us call each of these sets of simultaneous sense terms[1] a *view*.

(1) In what follows we shall only consider minimal terms, each of which is the appearance of a material point.

Relations of Position, Simultaneity, Qualitative and Local Resemblance

Each view is a space. The initial view presented the structure formed by the immediate relations of position of the sensible elements (reduced, for simplicity, to the connection of pairs). This structure can be expressed by the formula $G(v_o, N)$, where the function G denotes the set of axioms of point and congruence, v_o the initial view, which assumes the formal role of the class of points, and N the intuitive relation of connection of two pairs of sensible elements, which assumes the role of congruence. Now, *within each one of the views* which succeed the first, this same structure is found. The formula $G(v, N)$ holds for any view v. It expresses the fact that any view furnishes a complete interpretation, a new example, an original solution of the axioms: it says that any view is a space.

Resemblances among the elements of different views. Thus we are now confronted with a plurality of sensible aggregates of the same nature, the views, each of which exhibits the same structure of formal laws among its terms by aid of the same primitive nexus, i.e. the connection of pairs. But these successive totalities might have nothing else in common. The sense terms which comprise them might *not resemble each other in any respect.* Each view would then be entirely new, entirely original, by virtue of the quality of its elements.

However, we shall obviate this radical newness by granting our subject the perception of *local resemblance* and *qualitative resemblance* among data of different views (restricting ourselves throughout to the perfect resemblance of each type). He then compares the elements of any two views in two ways. On one hand, an element *a* of a view A resembles an element *b* of a view B qualitatively: we say that both of them are of precisely the same *shade.* On the other hand, this same element *a* of a view A locally resembles – and this is quite a different kind of resemblance – a second element *b′* of the same view B: we say that both of them occupy exactly the same place in the visual field. These two kinds of resemblance, extending from the elements of past views to those of the present view, moderate its newness and turn the succession of views into something like the various scenes of a play in which the same actors perform on the same stage.

Finally, in order to simplify the problem as much as possible, we shall exclude indiscernibles as we did in the previous examples.

Let us suppose that of the material points filling the space, all of which the subject perceives distinctly at the same time, no two have the same shade. Thus the appearance of the same luminous point in all the views will be united by the simple relation of perfect qualitative resemblance, so removing the particular complexity which arises from the fact that, in our world, two of the same object do not always resemble each other, and two similar appearances do not always belong to the same object.

Views, sensible positions, objects. Let us for a moment disregard the relation N of connection of two pairs of terms of the same view in order to consider *local resemblance* (L say), *qualitative resemblance* (Q say), and simultaneity which may be regarded as *temporal resemblance* (T say).

The sensible elements are thus classed in three ways. Any element x belongs to the class of elements which resemble it *temporally*, to the class of elements which resemble it *locally*, and to the class of elements which resemble it *qualitatively*. We have already called the first of these classes the *view of x*. The second will be called the *sensible position of x*, and the third the *object of x*. (In this we are not making any presuppositions about the existence or non-existence of simple entities having a more natural right to these names: we have no need for them, cf. Part Two, Chapter II.) We shall further call the members of an object its *appearances*, and when a sensible position, an object, and a view have one element in common we shall say that this *object appears in this sensible position in this view.*

Views, sensible positions, objects – these are the three ways of classing the sense data of our imaginary spectator through resemblance. For him the universe is the content of all the views. It is also the content of all the sensible positions. It is equally well the content of all the objects; for these are nothing more than three ways of dividing the fundamental aggregate of all sensible elements with respect to three different resemblances.

Fundamental laws. The three relations which are added to the original relation of connection when nature is no longer immutable will become the basis for a new scientific development. The properties they manifest, either in isolation or in combination with each other or with connection in fact furnish new laws.

Taken one by one, the three relations, local resemblance (L), qualitative resemblance (Q), and simultaneity or temporal resemblance (T) have the same characteristics of *transitivity* and *symmetry* which puts them in the formal class of perfect *resemblances*. It is by virtue of these two characteristics that the relations L, Q, T distribute their terms into the homogeneous classes we have called sensible positions, objects and views.

If we pass to the combinations of the resemblances L, Q, T, that is, to the conjunctions of the views, sensible positions, and objects, we notice the following double law which is an immediate consequence of our hypothesis: *Each view has one and only one element in common with any sensible position on the one hand, and with any object on the other.* Or again: *In any view, each sensible position is presented by one and only one datum, and each object possesses one and only one appearance.*

We proceed finally to the laws which relate the connection N with the resemblances L, T, Q.

The combination of N with the single simultaneity T has already furnished the following law: for any view v, we have G(v, N). This asserted that in each view the connections of pairs of elements obey the formal laws of congruences of points, or again, form a space. The combination of connection N with simultaneity T *and* local resemblance L or qualitative resemblance Q will express the relations of sensible locality or quality among these successive spaces.

Let us take local resemblances first. Let a, b, c, d, a', b', c', d', denote any sensible elements whatever; then we have

$$(ab)\ \mathrm{N}\ (cd).\ a'\mathrm{T}b'.\ a'\mathrm{T}c'.\ a'\mathrm{T}d'.\quad a\mathrm{L}a'.\ b\mathrm{L}b'.\ c\mathrm{L}c'.\ d\mathrm{L}d'\quad \text{implies } (a'b')\ \mathrm{N}\ (c'd')$$

or, in more intuitive terms: *The connection is transmitted from one view to the others by local resemblance.* Let us represent sensible elements by points, simultaneity T by a horizontal line, local similarity L by a line descending from left to right, qualitative resemblance Q by a line descending from right to left, and the connection N between two pairs of elements of one view by a double bracket. A view is then represented by the points of a horizontal line, a sensible position and an object by the points of oblique lines pointing in different directions. The transmission of connection by local resemblance is then illustrated by sliding the bracket N in the direction L (Fig. 8).

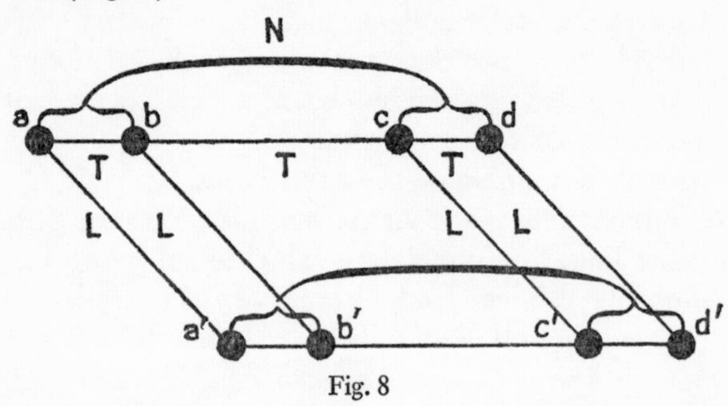

Fig. 8

But whatever the spectator has observed in connection with local resemblance, he also notices in connection with qualitative resemblance: like the pairs of sensible positions, the pairs of objects retain their connections in all views. (We must remember that we are considering a three-dimensional vision, in which the connection of two pairs of sensible elements expresses, not the equality of angular separation of the corresponding material points, which depends on the viewpoint, but their equality of distance which is invariant since we assume them to be motionless.) We therefore have a new law

$$(ab)\,\mathrm{N}\,(cd).\ a'\mathrm{T}b'\ .\ a'\mathrm{T}c'\ .\ a'\mathrm{T}d'\ .\ a\mathrm{Q}a'\ .\ b\mathrm{Q}b'\ .\ c\mathrm{Q}c'\ .\ d\,\mathrm{Q}d' \text{ implies } (\mathrm{a'b'})\,\mathrm{N}\,(\mathrm{c'd'})$$

in other words: the *connection is transmitted from one view to another by qualitative resemblance* (Fig. 9).

The transmission of the connection by local resemblance expresses the fact that the sensible field undergoes no deformation. The transmission of the connection by qualitative resemblance expresses the other fact that the set of material points perceived also remains invariant in its internal proportions. But it might also be possible that the set of material points contracts or expands *uniformly* with respect to the subject's sensible field. He would then observe the following. Two pairs of objects o_1o_2, o_3o_4, which in the one view v have two connected pairs of appearances e_1e_2, e_3e_4 have in any other view v′ two appearances $e'_1e'_2$, $e'_3e'_4$ which are also connected. But the pair of elements $e'_1e'_2$ which are the appearances of the objects o_1 and o_2 in view v′ *is not necessarily connected* to the pair $e''_1e''_2$ of elements of this view v′ which have the same sensible positions that the appearances e_1e_2 of the same objects o_1 and o_2 had in the first view v.

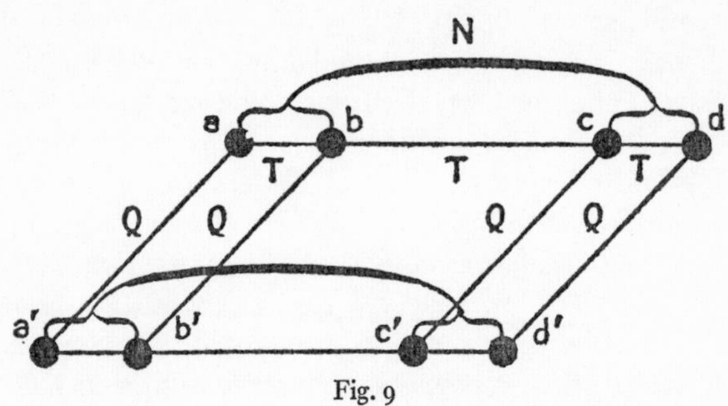

Fig. 9

Let us exclude any such contraction or expansion from our hypotheses, and imagine an absolutely unchanging perceptible world. The foregoing variation then never takes place, and we note the law

$$aTb \;.\; aLc \;.\; bLd \;.\; aQe \;.\; bQf \quad \text{implies } (cd)\,N\,(ef) \quad \text{(Fig. 10)}$$

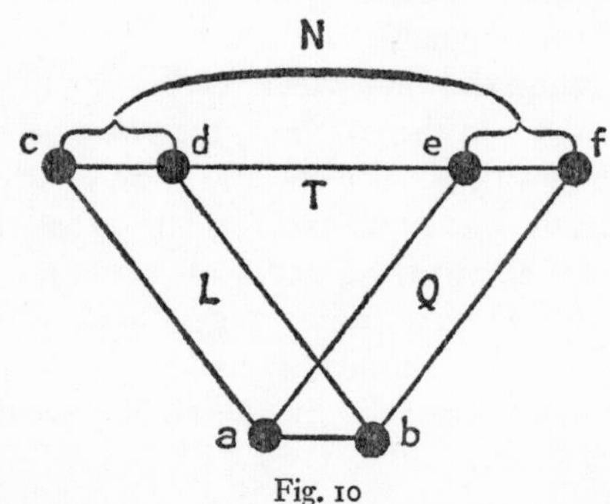

Fig. 10

This law differs from its two predecessors in one important respect. Whereas they merely state the ways in which a given connection in one view entails others in other views (either by local or qualitative resemblance), this law goes further: it says that a certain complex of the three resemblances – temporal, local, and qualitative – is sufficient to determine a connection. It, therefore, terminates in a connection without having started from a connection. This will enable us later on to relinquish the hypothesis which claims that the connection of two pairs of sensible elements is a primitive relation, so that we may instead *define* it to be precisely this complex of the three resemblances T, L, Q.

The science of nature is completely comprehended by the principles we have just seen. Let us summarize them: *Within each view, the sensible elements and their connections possess the properties of points and their congruences* (*law I*). *The elements of one view are in one to one correspondence with the elements of any other view, on the one hand through local resemblance, on the other through qualitative resemblance* (*law II*). *Further, the connection relation between two pairs* ab, cd *of elements of a view is transmitted to the two pairs* a′b′, c′d′ *of elements which correspond to them locally in any other view* (*law III*), *as well as to the two pairs* a″b″, c″d″ *which correspond to them qualitatively in any other view* (*law IV*). *Finally, two pairs* a′b′ a″b″ *of elements of a view which correspond, one locally, the other qualitatively, to the same pair* ab *of elements of another view are connected* (*law V*). Let us now see what structure, what form for the world, results from this.

The world is reduced to a single view, and at first we assumed that its only elements were the sense data which comprised it: above them we immediately found the view which, embracing them all, was the whole universe. Now, on the contrary, the sense data form various classes – sensible positions, objects and views – which are, with regard to the sum-total of experience, second-order elements, more complex and abstract than the sensible elements they contain. Let us consider the set of these positions, the set of these objects, the set of these views, each of which embraces the whole of the sensible world in its own way. What order does each of these three classes of classes present to the mind?

The space of sensible positions. Let us first of all consider the set of sensible positions; by which we understand, it is to be recalled, the classes of sense data formed by local resemblance. Consider a certain view v. Each sensible position has one element in this view; conversely, each element of this view has a sensible position. The subject can then think of any sensible position as the *sensible position of this*, and as *the set of sensible positions of the data which are here*. He can regard the set of elements of view v as a picture of the set of sensible positions, each element representing the sensible position of which it is a member.

Now these same elements, through the connections of their pairs, also form groups possessing the formal properties of equal distances, hence of straight lines, circles, spheres, and of all collections of points with which geometry is concerned. The mind might conceive the fancy of transferring this classification of elements of view v to the sensible positions they represent by classing together as 'connected' the pairs of positions represented by connected pairs of elements. To the straight lines, circles, spheres, formed by the elements which view v displays before the mind would then correspond other 'straight lines', other 'circles', other 'spheres' formed by the sensible positions these elements represent. The geometry of the elements of view v would be reflected in a geometry with more complex terms, in which *points* are expressed, not by the *elements of view* v as before, but by the *sensible positions* of which these elements are part, and in which congruence is no longer expressed by the *connection of pairs of elements of view* v, but by the relation among *pairs of sensible positions* which consists in being occupied by connected pairs of elements in view v.

Fancy, we said. In fact, the order of connections of elements of view v, now taken as representatives of their sensible positions, does not yet truly belong to these positions themselves. In an instant the view v will have been replaced by a view v′: will the pairs of sensible positions which are represented in view v by connected pairs of elements remain so in view v′? Varying with the view chosen, these 'connections' of sensible positions would not characterize these positions. With regard to these the connection would remain accidental, entirely lacking this indifference to its various members which exactly defines a property of a class.

What would be a necessary condition for the 'connections' of sensible positions defined by the connections of their members in a view chosen at random, to be an order characteristic of these localities themselves? We have just indicated it: it would be necessary for these relations to be *independent* of the choice of view. Now *this independence is precisely what the law of transmission of connection by local resemblance asserts.* The relation between two pairs of positions thus defined is in fact associated with these pairs themselves. But it faithfully reflects the connection of the pairs of elements of any view: and the latter in turn reflects the congruence of the pairs of points. Let l denote the class of sensible positions and N_l the relation between two pairs of these positions whose members in any view form two connected pairs. We then have

$$G(l, N_l)$$

that is to say, *the sensible positions and their relation* N_l *form the points and congruence relation of a space.*

The space of objects. The same clearly holds for *objects* or classes of sense data grouped by qualitative resemblance. On the one hand every object has, like every sensible position, a member or appearance in each view, and on the other by the connection of pairs of data is transmitted from one view to the others by qualitative resemblance as by local resemblance in virtue of law IV. Thus among pairs of objects there also exists a relation of 'connection' N_o consisting in the connection of pairs of appearances of these objects *in every view*, which exactly reproduces the formal characteristics of the data belonging to one view which are the formal characteristics of the congruence of points. Denoting the class of objects by o, we have

$$G(o, N_o)$$

that is to say, *the objects and their relation* N_o *form the points and the congruence relation of a space.* Thus, after the sense terms interior to each view and the sensible positions, the objects in turn illustrate geometry in their own way.

The space of a family of parallel views. It remains to examine the order of *views*. But here things are not so simple, and there is a slight shift in aspect. Again let us place ourselves, along with our spectator, in the presence of a certain view v: can we, right away, represent the set of preceding views in the intuitive order of this view v? Let us choose three objects o_1, o_2, o_3 which are actually presented to us through the three elements e_1, e_2, e_3 (not in a straight line), and let us ask at what other sensible positions in the preceding view v′ appearances of these same objects o_1, o_2, o_3 were found; then let us note down the elements e'_1, e'_2, e'_3 which have these localities in view v. In fact it follows from law V that if in two views three unaligned objects appear in the same positions, *all* the objects then appear in the same positions, and we regard two views of this sort as a single one[(1)]. Each preceding view v′ is then precisely representable in the view v by the three elements e'_1, e'_2, e'_3 which have the sensible positions in view v that the appearances of the three selected objects o_1, o_2, o_3 had in view v′.

But let us simplify still further. First of all let us suppose that all the subject's displacements are *translations*. An object then cannot appear in the same sensible position in two views without these two views being identical: each view is, therefore, adequately characterized by the sensible position of the appearance of one object, and not of three as before. Choose an object o: in the view v at present displayed before the mind, each view v′ will be represented by the element e' which indicates in v the sensible position of the appearance of the object o in view v′: *this will be the view in which this object appears in this position.* (As for the view v itself, it is clear that it is represented by the selected object's own appearance.) On the other hand, since the preceding views are always finite in number, it is far from being the case that each element of the view v represents one of these views.

But our observer here recognizes an essential difference which distinguishes the views from the sensible positions and the objects. From the outset of his experience he has been acquainted with all the positions and all the objects, which can subsequently only increase in new members. On the other hand, the views are each presented successively and in their entirety, and a new one is presented each instant. In thinking of the whole set of views he must therefore also consider future views. He cannot see them in advance, since we have not subjected his translations to any rule. But by

(1) This confusion of identical views alters the meaning of the term *view*: instead of designating a class of simultaneous data, it designates henceforth only the *content* of this class, *i.e.* the way in which it distributes qualities among the sensible positions.

relying on the analogy of views already seen, he can reject certain conceivable views as contrary to the laws of nature: such would be the case for a view in which an object appeared in the same position as in a known view, while another object appeared in a different position. The remaining views, not excluded by any law, constitute the *possible views*. We may represent them in the present view v in the same way as the past views, that is, by the elements of this view v whose position will be assumed by the appearance of the object o. Now it so happens that there are precisely as many possible views as there are elements in any given view, so that each element of the view v represents a view.

We are now in the same position with regard to views as we were previously with regard to positions and objects. We need only repeat the considerations which are already familiar to us. We can apply the classification of the elements of view v into connected pairs (and, as a result, into 'lines', 'circles', 'spheres', etc.) to the various possible views that these elements represent, by agreeing to say that two pairs of views are 'connected' when the elements of view v which represent them form two connected pairs.

But this relation would characterize two pairs of views only if it showed itself to be independent of both the view v and of the object o chosen as the basis for representing the views.

Now this is precisely what happens[1]. Let us then form the relation N_v between two pairs of views which consists in the fact that,

(1) For the view this results from the transmission of connection by local resemblance (*law III*) and for the object, from the new law

$$aTb \,.\, aTa' \,.\, aTb' \,.\, aLa'' \,.\, b'Lb'' \,.\, aQa'' \,.\, bQb'' \,.\, a''Tb''$$
implies $(aa')\ N\ (bb')$

which expresses the restriction of the observer's movements to translations, by positing that the appearances of all objects are displaced through the same distance in passing from any view to another (*law VI*) (Fig. 11).

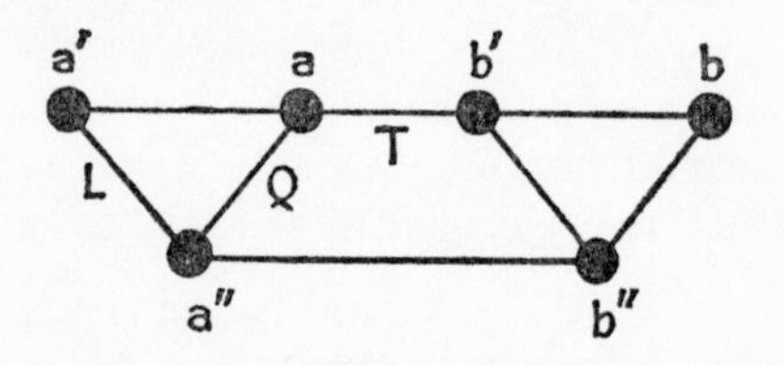

Fig. 11

for any object, the two pairs of sensible positions of its appearances in these views are in the relation N_v (for this is what it amounts to), or, to put it differently, in the fact that the two pairs of views displace the appearances of all objects by the same distance. Let us call the set of views v_1: we have

$$G(v_1, N_v)$$

that is to say, *the views and their relation* N_v *form the points and the congruence relation of a space.*

Formation of a total space of views. Up to this point we have only considered views which are all oriented in the same way. We now lift this restriction. Let us introduce views of any orientation whatever into the spectator's experience by subjecting his body to displacements which are no longer merely translations. The views he regards as possible then form a multitude of families v_1, v_2, ... of views oriented in the same way. Each one of these families can easily be defined by the persistence within it of law VI (equal displacements of the appearances of all objects). Each of them forms a space, since we have $G(v_1, N_v)$, $G(v_2, N_v)$, etc.

All these spaces of views manifest a close relationship which enables us to collect them into a single space. For it is possible to divide the totality of views into classes such that each class contains exactly one view of each space of views, and in all these spaces the geometric relations among views in the same classes are the same. In intuitive terms, it is possible to apply these spaces of views on to one another in such a way that their order structure coincides. Indeed, the spectator, after selecting an object o, can collect together those views in which this object appears in the same sensible position: the classes of views thus constructed have the required property. If we form the relation N_t between two pairs of these classes which consists in the fact that the two pairs of views belonging to them in any space of parallel views are in the relation N_v, these classes in turn constitute the points of a space whose relation N_t is congruence and in which any possible view is situated. Let t be the collection of these classes: once again we have $G(t, N_t)$.

But this total space of views suffers from a defect that we have already noticed in the total space of motions in a previous chapter:

its construction is merely arbitrary. It depends on the selection of an object o, which cannot be guided by reason. If we change objects, the classes of views forming its points are decomposed and reformed in a different way. Thus, this total space of views has no natural existence for such an observer(1). We shall observe quite a different state of affairs when we restrict his vision to two dimensions only.

(1) This same type of indeterminacy of the total space is found again in mechanics, where it corresponds to the special theory of relativity.

CHAPTER VII

Reflections on the Preceding Universe

In such a visual world, geometry is completely expressed many times. It is fully illustrated in the first view seen: its *points* are the *elementary data of this view*, its *congruence* is the basic relation N which we have called *connection of pairs of these data*. Then, it is again illustrated in each of the subsequent views, successively taking as points and congruence the elementary data of each view and the connection of pairs of them. All these interpretations, as numerous as the views, may be arranged into a primary class: for as congruence they have the same relation N, and as points terms all of the same nature. Geometry is next illustrated in the set l of *classes* of sense data grouped by local resemblance, or sensible position. As points it then takes these classes, and as congruence, the relation N_l between two pairs of sensible positions whose members in any view form two connected pairs. Geometry is again illustrated in the set o of classes of sense data grouped by qualitative resemblance, i.e. objects, taking as points these new classes, and as congruence the relation N_o between two pairs of objects whose members in any view form two connected pairs. Geometry is also illustrated in each set v_1 of identically oriented possible views (that is to say, a set of views linked by law VI): as points it takes these views, and as congruence it takes the relation N_v between two of their pairs in which the appearances of any object have sensible positions whose pairs are in the relation N_l. Finally, geometry is illustrated in each total set t_1 consisting of the classes of views in which the members of the object o_1 are in the same sensible position: the points are then these classes, and congruence is the relation N_t of two pairs of them which 'cut' any space v of parallel views in two pairs of views which are in the relation N_v.

This visual universe therefore furnishes the axiom system G of point and congruence with the following solutions: *for any view* v, we have $G(v,N)$; $G(l,N_l)$, $G(o,N_o)$; *for any set* v, $G(v,N_v)$, *and for any set* t, $G(t,N_t)$. The structure G is first presented as the structure of the simple connection relation N within a view. It is then extended to the whole universe, regarded as the set of sensible positions as the set of objects, and finally as the set of possible views. The plurality and diversity of these sensible spaces would be, to the fictitious being we have imagined, the subject matter of philosophy.

Irreducibility of these various geometric orders to one single fact. He would observe that they differ in epistemological status. The space interior to each view furnishes an intuitive order without succession. On the other hand, the space of objects and the spaces of views include, among the relations N_o and N_v which order them, the comparison of successive views. They are founded on laws IV and V which say that the perceived material points remain motionless: laws of physical nature devoid of intrinsic self-evidence. If the order: *for any view* v, G(v, N) can be to our subject a geometry in the old sense of intrinsic self-evidence, the order $G(o, N_o)$ and $G(v, N_v)$ are, to him, nothing more than a purely inductive physics[1].

We often feel that the geometric structure of nature has elements of both the intuitive and the elaborated; for we find geometric structure both in the arrangement of the parts of an instantaneous perception and in the comparison of successive perceptions. But since this type of intuition is very imperfect in us, it is natural to seek in its very imperfection the reason why the geometric structure of the universe for us includes a share of succession and a share of arrangement. However, we have just defined and posited a perfect spatial intuition, and other spaces which are founded on the laws of succession, have appeared all the more clearly. By setting aside the source of additional complications, namely the imperfections of our senses, we have more clearly exposed the fundamental plurality of the spaces which order, in the simplest conceivable visual universe, the data of a view, the sensible positions, the objects, and the views themselves.

(1) As for the order of sensible positions $G(l, N_l)$, it would be successive and inductive or, on the contrary, timeless and intuitive, according to whether one rejects or adopts the hypothesis which analyzes the local resemblance of two successive sense data into an identity of 'occupancy' of the same simple entity, invariably present, which would be their individual sensible position. As a matter of fact, in this hypothesis the relation N of connection of pairs of sense data corresponds to a relation n among the individual sensible positions occupied, and the transmission of the relation N in one new to succeeding views by local resemblance, which is the foundation of this order $G(l, N_l)$, ceases to be an inductive law of heredity and becomes instead a logical indentity.

But how can we refrain from conjecturing that all these spaces are derived from the simplest and only intuitive one among them which would already include all the geometry of nature? In its desire for unity, the mind attempts to reason as follows: all nature reduces to sense data; now any sense datum enters immediately into the intuitive geometry within perception; hence the same holds for any collection of sense data; therefore, it is certain that objects and views will in turn be ordered geometrically.

But this is a highly confused train of thought. For the class of data which form any object or view pervades the sensible field and fills it completely, so that the objects or views, being omnipresent in it, derive no order whatsoever from the mere fact that they are there.

Or again, using a more indirect argument we might say: if I try to imagine the set of objects or the set of possible views, I only succeed with great difficulty, or even do not succeed at all, if I fail to conceive them as the intuitive set of heres and theres of the sensible field; consequently, the order of this field cannot fail to be communicated to them in my thought.

But we know that even if the premise of this argument were true, its conclusion would not follow. Indeed, it presupposes that, to the mind, any order among symbols refers to things, whereas, on the contrary, this order may represent nothing to it. If, in the universe studied, the objects and views do in fact possess a geometric order analogous to that of the parts of the same view and are to this extent symbolizable by it, the reason does not lie in the control over the imagination exercised by this first intuitive order, but solely in the particular laws governing the succession of views, inductive laws which are devoid of self-evidence.

Thus, in a sensible nature restricted to vision and simplified in the extreme, the plurality of spaces already has the last word. The imagined subject cannot, in any valid way, discover a supreme space but only intermingled spaces whose terms, order relations, and connections with time and intuition are all distinct.

The spaces of views. Of the four kinds of space our spectator knows, the space of a view, and even the space of sensible positions, are structures which may be called intuitive and personal; the space of objects and the spaces of views, on the other hand, are empirical and impersonal structures, provided we do not ask too much of the rather vague terms 'personal' and 'impersonal'. The spaces of views, which are perhaps less familiar to us than the others (for a reason we shall see presently), particularly merit our consideration. For the opposition between the order of views and the order of a view exposes the error (however small) of reasoning about the nature of spatial order as if this nature were unique.

What, in fact, is a point in a space of views? It is a complete view, that is to say, a complete content of the sensible field. It is a scene which is already a complete space in itself such as geometry requires. However, these successive spaces, through their resemblances and the differences of their contents, are in turn ordered as so many points in accordance with geometric form. Thus, the intuition which embraces the points of a view within their order of *heres* and *theres* is at each instant contained in a single point of a space of views: the order of these points could not escape it more completely.

The geometry of nature has two contrary relationships with intuition. One is in the immediate classification of the interior details of an instantaneous perception. But here another is found – and not by consequence – in the classification reflected by each perception taken as a whole, within the set of possible perceptions, of incompatible aspects of the world, each of which without exception is sufficient to pervade the sensory extension in all its plentitude and, consequently, the imagination. No longer interior to the total perception of an instant, geometric order is applied to this total perception itself, so that it may in its turn be placed in the set (which can no longer be intuitive) of all total perceptions. Thus any universe in which there exists a space of views contradicts the philosophy which sees 'space' as a mere structure within each view.

Likewise, the consideration of spaces of views appears fatal to the thesis of the subjectivity of 'space', which is usually associated with that of its intuitiveness. A space of views in fact orders not only the successive perceptions of the same subject, but also the simultaneous perceptions of different subjects.

Let us imagine that our character can penetrate directly into the consciousness of other beings like himself perceiving the same universe. He would become acquainted with the views present to these other minds, views different from the one he actually at present perceives, but similar to certain of the views he regards as possible. In the geometry of possible views he would then place the perceptions of his companions as if they were his very own. Thus, three of these perceptions could be in a straight line, or else on a circumference of a circle with a fourth for the centre. This geometry of possible views has therefore provided the architecture of a common world in which the individuals' various views are co-ordinated by virtue of their systematic differences.

We may observe here that Kant, by maintaining that space is subjective (and undoubtedly having before his mind some idealized notion of an intuitive and personal space analogous to the perfectly geometric sensible field of this example), has sensed the danger of ending up with as many spaces as subjects. He realized strongly, although obscurely, that there is a unique space for all of us. But he did not dream of distinguishing several spaces of opposing characters. He relates this feeling he has of the existence of the same space for all men to this space immediately apprehended by each of us, whose subjective character he imagines he can clearly see. What a strange, almost mystical use he then makes of the notion of a non-individual subjective! He seems to want us to think: 'Space is posited by a pure intuition of the subject, but this intuition reflects none of the characteristics of the empirical me: as subject it has, not Tom, Dick or Harry, but only the man in them: the intuition is thus a unity, and space, therefore, is one for all men.' A false conclusion: for all that is proved is the *perfect resemblance* of your intuitive space and mine, and not their *numerical identity*: since space is associated with the subject's activity, there are two spaces just as we are two subjects, and there is no going back on it.

Now what establishes geometric order among the views apprehended by the various subjects, each of which fills his personal space, is not yet the perfect resemblance of these spaces, but the systematic *differences* of the perceptions they have as contents. In order for the views of all to be co-ordinated in a single spatial order, it is not sufficient that they have *a priori* similar structures: it is in addition necessary for them to have different sensible contents – different according to the complicated and precise rules of perspective. It is these differences between your perception and mine, each in its own immediate space, that constitute the single space in which both of us are situated. That the existence of these regulated *variations* among the sensible *contents* simultaneously present to the various subjects results from the fact that their minds are *similar* and that the *structure* of space is the same for all – is not the error manifest?

Leibniz, father of the idea of a geometry of views of the universe, has seen more clearly than anyone the double nature of spatial order, personal and subjective in one form, impersonal and universal in another. Indeed, according to him spatial relations sometimes appear as if they connected the simultaneous objects of each monad's perception, and these same relations do not hold among the monads themselves. But the monads have an analogous order nonetheless, since each possesses, at a given moment and by virtue of its complete perception, a characteristic *viewpoint*. The duality of the space of a view and the space of views shows itself clearly here.

Independence of the order of views and the order of objects. There is still another question which will prove instructive. Of the two empirical and impersonal forms of geometric order of nature in this example, the order of objects and the order of views, why is the second less familiar to us? This arises from the fact that in practice our vision distinguishes only two dimensions. In fact the result of this limitation is the association of any possible view with a certain object through a single relation which we express by saying that this view is *what is seen from this object* – for example from the top of this tower, from the end of this telescope. Later we shall investigate the nature of this relation between view and an object. But it is so natural to us that we always conceive the geometric order of views as being merely that of the objects in which they are situated. Let us ask a physicist what he means by the viewpoint of a certain observation of the universe. He will probably reply in the following way: 'By the viewpoint of an observation I mean the position of the observer's body in the totality of all other bodies.'

But in this universe, such an answer is impossible. For none of the views has a distinctive relation to any of the objects. Our subject cannot, therefore, refer to any view whatever as 'the view related to this object'. In spite of this, a geometry of views emerges, thanks to which each view takes its place in the totality of the world, regarded not as the totality of objects but purely and directly as the totality of views. Thus, the first visual universe demonstrates that a geometric order of the views of the world can exist without these views being located here or there in the totality of bodies. This possibility is important, for even in a world where views are closely linked to bodies, it would be possible for the order of views to be the more fundamental of the two.

CHAPTER VIII

Elimination of Relations of Position

In the preceding universe, we started from an intuitive order of data within each view: then, the same kind of order was discovered in the relations of sensible positions in those of objects and in those of views. But each of the relations N_l, N_o, N_v ordering these various totalities in the manner of geometric congruence contains in its definition the initial intuitive relation N among the data of a view. It is, therefore, an *extension* to positions, objects, and views, of an intuitive order in each view, which remains the central element – not sufficient, but necessary.

Let us now suppress or disregard this order: we shall see that the others continue to exist, and that their dependence on it is, therefore, only apparent: besides, it might be nothing more than an illusion itself.

Let us in fact return to the three instances of this order we have given. The alignment of three stars, the equal separation of two pairs of stars, in this we felt we could perceive a simple basic detail in the spectacle of the heavens. But on the contrary, it is possible that all relations of this sort among various parts of a view are composites, composed of relations which actually compare the present view with previous views.

Imagine two observers, one perceiving the relations of position among the data of each of his views, and the other not perceiving them. For the first, such pairs, such triads of data of a view present a direct resemblance (equality of separation, property of alignment). The second, on the other hand, does not discern any particular relationship among these data. For him, any two pairs of triads resemble each other or differ in the same way. He cannot understand the classification of pairs into equal or unequal pairs or of triads into aligned or unaligned triads established by his companion. He is like a man who hears sounds perfectly and recognizes them with accuracy, but who is tone-deaf and senses no difference whatsoever between groups of three sounds forming a perfect chord and groups of any three other sounds.

But this tone-deaf man could discover a relation among the notes which the musician says form a perfect chord: he could even anticipate unfailingly the latter's judgement on any three notes heard for the first time, by counting their vibration-numbers, in order to apprehend their relationship. Likewise, our observer who cannot perceive the resemblance of figures can learn, and more readily still, to distinguish the figures which his companion would declare to be similar.

It is enough for him to observe the operations the other performs for the purpose of *verifying* the exactness of his immediate judgements. To assure himself that three stars really appear to be in alignment, he observes that they lie along the edge of a pencil, and to verify that this edge itself is a suitable standard, he observes that it reduces to a single point when viewed from one end. Thus we are again saying that three data *a*, *b*, *c* of a view are aligned if their sensible positions A, B, C can be occupied by the simultaneous appearances of three small objects α, β, γ such that there exists a view in which the appearances of two of these objects are both hidden by the appearance of the third. Now this is a relation constructed out of local resemblances, qualitative resemblances, simultaneities, and successions which the second observer can search for and grasp.

Similarly, his companion verifies the equality of separation of two pairs of stars by successively covering both with two fixed points on a pencil held at arm's length. If his visual memory were perfect, it would even be enough for him to turn his head and bring the second pair of details into the sensible positions occupied by the first pair: if the operation is successful, the equality of separation is verified. But this is, once again, an experiential relation which the second observer can grasp as well as the first.

It is not even necessary to perform any movements: the experiment can have been made once for all. Indeed, if three data of a view are aligned, so also are the three data of any other view having the same sensible positions; equality of separation enjoys the same property. Hence the positions of the visual field can be experientially classified once for all according to their alignments and equalities of separation. Our second spectator could, therefore, learn to judge, in the presence of a new view, the resemblances of the figures found there, not only as well, but also as immediately, as the other.

If we meet these two observers after the end of this apprenticeship, we shall not be able to discover any difference between them, for, by seeing them relate the same figures by means of the same words, how can we guess that they apprehend very different relations among them? Perhaps in fact they themselves would not suspect this unless they explain themselves extremely precisely.

'By alignment and equality of separation of certain data of a view', the one would say, 'I understand relations of simple character which do not refer to anything beyond the present'. The other would answer: 'I do not know how to take what you say, for I myself understand by these words very complex relations, constructed from qualitative and local resemblances, from simultaneities and successions, relating present data through the intermediary of past events whose scene they occupy. Far from being a direct resemblance, the relation of two equal pairs, of two triads in alignment, is to me but a relationship consisting in certain past facts, just like human relationships. But are you not deluding yourself? Your actions have taught me the experiential meaning of these two expressions: Are you certain that originally you did not learn it in the same way, only to forget it now that the necessary experiments are performed, and you know by heart what sets of sensible positions they relate?'

'It really seems to me', the first observer would say, 'that among these sets I apprehend direct relations with no historical content. But they are invariably duplicated by the indirect relations you refer to. I even use these relations alone when I demand great precision (for instance, when I wish to construct a map of the sky), because the observations which constitute them are more precise than my simple perception of alignment or equality.'

'Let us then agree,' the other would reply, 'that this simple perception not possessed by me which you think you have is not very useful. It adds to the picturesqueness of the world without adding to its order, since other relations which you yourself say are more precise yield the same classification of appearances.'

Which of these two observers do we most resemble? We seem to have some basic feeling for the similarity of figures, but we trust only indirect comparisons through real or apparent displacements of objects. The intuitive relations of position, if they exist, are therefore not essential to the scientific order of nature: we can already see this in the simplified nature just analyzed.

In this example, all the relations of position are represented by the relation N of intuitive connection among the pairs of data resulting from congruent pairs of material points. Now this immediate relation is duplicated by a complex relation composed of qualitative, local, and temporal resemblances. In fact, law V states that two pairs of data ab, $a'b'$ of a view are connected whenever a and b have the same sensible positions, and $a'b'$ the same objects, as a pair of data $a''b''$ of some other view. Conversely, no law precludes the existence of a view in which the positions of ab are assumed by appearances $a''b''$ of the objects of $a'b'$, provided exactly that the pairs ab, $a'b'$ are connected (Fig. 10). In the set of possible views, this complex relation, $\mathcal{N}$ say, is therefore equivalent to the simple relation N since it related the same pairs of data.

Thus let us deprive our subject of the perception of relation N and of any other relation of position among various sets of data: in his universe he nonetheless finds the multiple geometric structure we have analyzed, in which $\mathcal{N}$ takes the place of N. True, it takes him longer to do so; for instead of connecting these same pairs of data by means of a manifest relation, he has to wait for favourable experiences which enable him to compare them. It is equally true that he is no longer aware of any simple version of geometry in nature. All the significations of congruence – and of any other geometric relation – are now complex relations fashioned from local, qualitative and temporal resemblances, involving several successive perceptions.

CHAPTER IX

The Geometry of Perspectives[1]

Leaving all relations of position to one side, we shall confine our attention to local, qualitative, and temporal resemblances. But let us inch one degree closer to the nature we know: let us study the universe of a visual sense for which the diversity of sensible positions no longer corresponds to the diversity of the positions of bodies perceived, but only to that of their directions. This is in fact the so-called *distant* vision, and it is the only one precise enough for science. For if it is true that we sense a difference between two visual data which are the appearances of bodies lying in the same direction, but at different distances, when these bodies are close by, this sensation is far too vague to be employed in observations[(2)].

Therefore, let us show what idea of the geometric order of the world an observer would possess were he reduced to a two-dimensional sense of vision; to him life would be like a moving picture show, or even a magic lantern performance, since we are obliterating all perception during displacements. This lesser differentiation of the order of data in each view gives rise to certain new features, essential to our universe, which return us to a very simple foundation.

We can no longer postulate a plenum of visible matter, for a two-dimensional vision would be blind to it. It requires transparent spaces. Let us, therefore, assume six material points, resembling six stars of different shades in a dark sky. These six stars are not in the same plane, and three of them are collinear: this last assumption, which simplifies many things, may perhaps not be indispensable.

The observer to whom we present this sight of six stars by successively and randomly placing the material point which is his

(1) Cf. B. Russell, *Our Knowledge of the External World*, Chap. III.

(2) The thesis which states that visual data would all be 'at a distance', but without *differing* in this respect appears to be confused. In any case, assuming the identity of data resulting from bodies at all distances lying in the same direction, it in no way removes the two-dimensional character of the visual manifold.

body in all sorts of positions, perceives, like his predecessor, local, qualitative and temporal resemblances among his sense data, which are therefore again classified into *sensible positions*, *objects*, and *views*.

The three objects A, B, C which correspond to the collinear stars manifest the following particular relationship: the appearances of two of them are both *absent* from certain views. Let us call these views (ABC) views: we shall say that two of these objects are hidden by the third, or that these views are *sighted* along the straight line ABC. When the three objects thus related all appear together, they can be found only in certain triads of sensible positions. We shall call these triads *aligned*, and *alignment* the class of positions aligned with two given positions or with two positions aligned with the latter. An alignment is the possible appearance of a plane for an observer situated in this plane and endowed with a sense of vision which surveys all directions at once: it is a circle of horizon. Indeed, if we compare the visual field to a sphere, alignments are great circles. This sphere idea is useful: we shall make use of it in our figures. But it goes without saying that a sensible field, not being an object, could not possess a form we can trace or model.

Consider two views V, V_1, separated by a displacement without translation of the observer. When we pass from one to the other, the appearance of objects slide along alignments converging in two sensible positions a,a' (one of which is the point in the horizon towards which we advance and the other the opposite point from which we retreat). We shall say that two views related in this way are *transformable into each other with respect to* a, a'.

Leaving the observer at the viewpoint of V, let us meanwhile make him execute a semi-rotation on the spot about the line joining this viewpoint to that of V_1: the new view is still transformable into V_1 with respect to the same positions a,a'. In fact, all the appearances have simply revolved through a semi-circle about $a\,a'$: they are, therefore, found on the same alignments intersecting in these two positions. Thus, two transformable views have orientations which are either identical, or separated by a semi-rotation about the straight line joining their viewpoints.

In this last case, any view which is transformable into both at once must, therefore, lie on the same straight line, so that the transformation is effected with respect to the same positions *a*,*a′*. The identical orientation of three views of which at least one and at most two are (ABC) views – we assume this so as to ensure that they do not all have the same viewpoint, a case in which the following argument does not apply – is then expressed by the condition that they be transformable two by two with respect to three different pairs of sensible positions: this is how we shall define a *family of parallel views*.

But first, this possible relative double orientation of the transformable views will furnish a definition of symmetry, of equidistance, and finally of congruence of sensible positions.

Let V be an (ABC) view which is transformable into a non-(ABC) view V_1 with respect to the pair of positions *m*,*n*. There is exactly one other view V′ which also satisfies this double condition: it is the view having the same viewpoint as V and separated from V by a semi-rotation about the line VV_1 (Fig. 12).

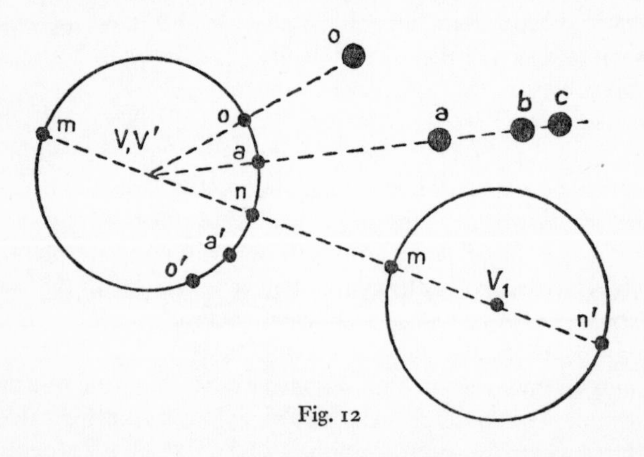

Fig. 12

The two (ABC) views V and V′ thus related through their common transformability into some non- (ABC) view with respect to the same pair of sensible positions *m*,*n* are therefore symmetric in *m*,*n*: the positions *o*,*o′* occupied in these two views by the appearance of the same object O will be said to be *symmetric with respect to m and n*.

All the pairs of positions *xy* such that *x* and *y* are symmetric with respect to *m* and *n*, *m* and *n* being conversely symmetric with respect to x and y, are part of the same alignment, which we shall call the *equation of m and n* (Fig. 13).

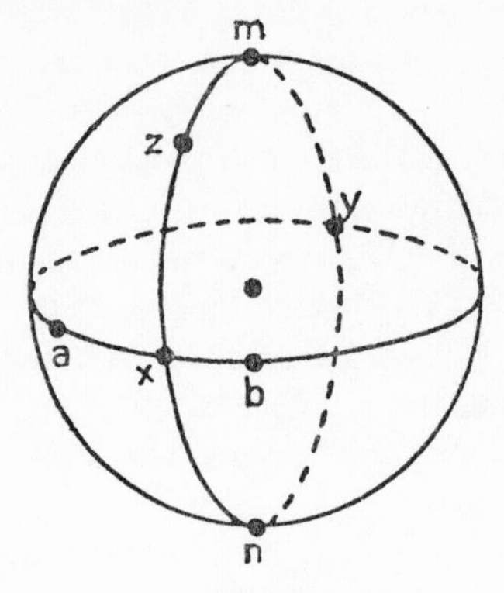

Fig. 13

Consider an alignment through m and n, cutting their equator in x: any position z in this alignment will be said to be *equidistant* from any pair a,b of positions on this equator which are symmetric with respect to x. Finally, if any intermediate term of a sequence of positions $pq\ x_1x_2,\ldots,\ x_nrs$ is equidistant from its two neighbours, the outermost pairs pq, rs will be called *congruent*.

The properties of these alignments and congruences completely determine the order of the set of our observer's sensible positions or, in other words, the structure of his field of perception. But this structure is no longer euclidean: the alignments of sensible positions possess only the order of the great circles on a sphere; the relation we have just termed their congruence only illustrates the congruences peculiar to the points of a spherical surface. If we continue to use the term *space* for a set of terms and relations satisfying the group of euclidean axioms, we must say that the set of sensible positions no longer forms a space.

The set of objects has also lost this property. Whereas the preceding universe provided as many objects as the points required by geometry, this one only possesses six, and a set of six terms clearly does not form a space.

What then is a space in this universe? What terms have the profusion of geometric points, what connections have the properties of its relations? Must we conclude that the data are too reduced, and that a two-dimensional visual experience is not sufficient to illustrate the whole structure of a space? No indeed, for if the space of sensible positions and the space of objects have vanished, the spaces of views have remained.

Let us in fact consider the family of views parallel to a given view, according to the definition we formulated a while ago. They correspond precisely to what we call the points of space: for to each point corresponds that view which represents the six stars seen from this point with the common orientation. We shall show that certain relations among these views reflect in our observer's experience the relations of rectilinearity or of distance between their viewpoints.

If a view X is transformable into two views A and B with respect to the same positions m,n we shall say that it is part of the straight line AB.

If a view X, common to three straight lines d,d',d'' is transformable into views of d with respect to m,n, into views of d' with respect to m',n', and into views of d'' with respect to m'',n'', and one of the positions m,n is symmetric with one of the positions m'',n'' in respect of one of the positions m',n', we shall say that the straight line d' is the *bisector* of the two others.

If, in this definition, we identify d'' with d, m'' and n'' with m and n, we shall say that the two straight lines are *perpendicular*.

If the four views A,B,C,X are such that the straight line AX is the bisector of the straight lines AB and AC and perpendicular to the straight line BC, we shall say that the view A is *equidistant* from the views B and C. Finally, if any intermediate term of the sequence of views $PQX_1X_2, \ldots, X_nRS$ is equidistant from its two neighbours, we shall say that the outermost pairs of views PQ,RS are *congruent*.

Obviously the relation between two pairs of views thus defined possesses all the properties of euclidean congruence, since in the universe considered it expresses the equality of distance of corresponding viewpoints. It is remarkable that the geometric order of views survives the disappearance of the geometric order of sensible positions and objects: the point is that, in order to represent space in experience, it is not necessary to associate with each point a sensible position or a thing, but merely an aspect of the world.

But up to now we have only considered a single family of parallel views: there are others, and each of them forms a space whose points are these views. The set of views therefore forms a multitude of spaces, and as yet we cannot tell whether all these spaces comprise a unity. The universe in the preceding chapter, constructed from three-dimensional views, has already presented this state of affairs. However, a new fact arises.

The space of 'points of views'. Before, we were able to superpose the various spaces of parallel views by selecting a certain object and then grouping the views according to the sensible position occupied by its appearance, classing together all the views in which this position is the same. The classes thus formed contain in themselves all the views: each family of parallel views is distributed among them at the rate of one view per class, and the geometric relations within these families correspond to each other exactly: if, in a family of parallel views, the views AB,CD belonging to the classes α, β, γ, δ, are congruent, then the same holds in the other families. We have, therefore, been able to regard these classes of views as forming a total space which derives its order from the intermingled orders of all the families of parallel views.

However, this total space was only constructed in an arbitrary way. Indeed, the classes of views that it has as points must be defined by means of a certain object, and, if the object is changed, these classes are no longer the same: for two views which appear in the same class if we consider the sensible position of the appearance of an object A, belong to different classes when we adopt the position of the appearance of an object B as our basis for classification. Thus the total space we could construct by superposition of spaces of parallel views had no natural existence in so far as it depended on a random choice.

This is exactly the limitation that disappears in a universe of only two-dimensional views. Here, the classes which unite the views of all orientations so as to furnish the points of a total space are formed in a unique way through a relation free of any arbitrary reference. A three-dimensional view has no more of a *point of views* than a statue: a two-dimensional view, on the other hand, has its point of views just like a picture, and the views of all orientations whose points of views are the same are linked by a general relationship among their contents: in all these views, the visual angles separating the objects remain constant. To each of the points of views at which we can place our observer's body in what we call empty space, something corresponds in his universe: a class of views in which the appearance of any two objects occupy congruent pairs of sensible positions. Let us call such a class of views a *point of views* – if we admit this kind of play on words for a moment.

As in the previous case, these classes include all the views: each space of parallel views is distributed among them, and the geometric relations of all these spaces are superposed, engendering a total space. But whereas before the classes in question were not well defined because the random selection of an individual object made their construction artificial, the collecting of views into 'points of views' is, on the contrary, unique and based on a general relationship.

Our observer for whom the views alone form spaces will see in the order of 'points of views', the most comprehensive space which contains and links all the others. For him it will be the geometric structure of the world, the space which orders the whole realm of the sensible.

But in addition to the formation of a unique total space embracing the whole set of views, the hypothesis of a vision reduced to two dimensions introduces a second feature characteristic of our world: it associates each visible object with a particular 'point of views'.

Association of objects with 'points of views'. In the case of three-dimensional views, we have noticed the absence of a non-arbitrary connection between the set of objects and the set of views: no view or class of views was associated with one object rather than any other by means of a distinctive relation. The objects formed a space, the views did too, but these two sorts of space were not connected with each other. Now, on the contrary, the objects no longer form a space – there are not enough of them for that – but, on the other hand, they are located in the space formed by the views.

In general, any view is fixed by the positions of the appearances of four objects: the position occupied by the appearance of a fifth object is then determined, and it is possible for our observer to calculate it on the basis of other views. Moreover, this determination is *con-*

tinuous: if the first four positions undergo a small variation, the variation of the fifth is equally small. Thus, when the appearances of four objects in several series of views approach four given localities by different routes, the appearance of the fifth object tends towards the same position in all these series.

However, certain objects are exceptions in certain possible views. We cannot calculate the position their appearance must assume. All methods of construction fail because of a particular disposition of the positions which constitute the data: the alignments whose intersection should fix the desired position all become indeterminate. Furthermore, the most complete discontinuity appears. According to whether we consider such or such a series of views in which the appearances of four objects tend towards the same positions by different routes, we see the fifth object in turn tending towards all the positions of the sensible field. The view determined by the positions of the appearances of the four normal objects is, therefore, a *singular* view with regard to the fifth: it so happens that all the views belonging to the same 'point of views', and these views only, share this singularity. Thus the observer is led to regard such a 'point of possible views' as singular with respect to such an object. Moreover, each of his objects has its singular 'point of views'.

What then is the special relation which in this way associates a certain collection of possible views with any object? We express it by saying that these views form the *vision of the universe viewed from this object*. In fact, when we move the material point which represents our observer's body towards the material point A along different routes, the appearances of five distinct material points tend towards the same group of positions, whereas on the contrary the appearance of point A remains fixed in various positions.

It is a remarkable fact that we manage, in a purely visual and completely unchanging universe, to associate with an object the class of views which we call the views perceptible from this object. For it was by no means obvious that the localization of a certain possible perception in a body admitted a signification independent of all causality. Might it not seem that a perceptual content was situated in a certain body – as for example, a certain view of the sky in a certain observatory – only through this body's relation to a particular body called the body of the observer, which itself determines, by concomitant variations, the views presented to its master? But in this world the observer is a pure mentality in his own eyes. He is not aware of having any body himself; he sees nothing more than chance in the succession of views. In spite of this, with each visible object he associates a class of singular views. If he could penetrate into another mind similar to his own, and if in it he found one of these views, he might say to himself: this mind is situated in this object – not, of course, with the full import we ascribe to these words in our richer world, but with a determined import nonetheless.

The mere substitution of two-dimensional views for the three-dimensional views at first studied, under the assumption that it prescribes empty spaces in visible matter, thus has important consequences. The space of sensible positions and the space of objects vanish. On the other hand, the spaces of views survive by becoming more complex, and come to be fused in the total space of 'points of views'. Finally, the objects which remain and no longer form a space are themselves insinuated into the space of 'points of views'.

The order of views thus becomes the unique space, the fundamental space of nature. At first we appear to have things the wrong way round. For we are accustomed to regard the order of the space of objects as fundamental, and to picture the order of the space of views as derived from the order of the objects in which they are physically situated. To us this seems a positive way of thinking. Nevertheless, the inversion, which, in this simplified world, makes the order of views or total perceptual contents appear as autonomous, and as the network of the very order of the bodies, is necessary. In the light of rigorous analysis it in fact provides the only possible application of space geometry to the content of such an experience.

Introduction of new objects. The sensible geometry which has just been presented involves only six objects, three of which are collinear. But let us introduce into our observer's world a new group of six stars satisfying the same condition. This new group will furnish material for a geometry independent of the first. In fact it provides, *as if the first did not exist*, its own definition of alignment and of all the entities which enter into the formation of spaces of views and of one total space of 'points of views'. The coincidence of the two constructions thus obtained is therefore not a necessity, but a simple fact of existence. When he discovers the six new objects, the observer cannot know anything about the rules which govern the displacements of their appearances from one view to another. Will these rules be the same as for the first group of objects? The hypothesis is natural, but it remains a hypothesis. Far from being necessary, it is easily conceivable as false.

Perhaps the new objects, by means of the displacements of their appearances, impart to the views an order which is not at all geometric, or else an order of two or only one dimension: this is conceivable, for we know that the points of a complete space are capable of being arranged in a series. Perhaps they lay the foundations for an entirely different geometry of views, a non-euclidean geometry, for example. But suppose that the appearances of the objects in the new group are displaced according to the laws already known. The spaces of views, the total space defined by considering only the appearances then have the same structure as the analogous constructions in terms of the appearances of the initial group. However, it does not yet follow that the order of views is the same. Indeed, these formally similar constructions are independent of each other: three sensible positions aligned with respect to the first group of objects might not be so with respect to the second: two views parallel in the one case might not be so in the other; two views might be of the same 'point of views' with respect to the old objects but of different 'points of views' with respect to the new objects. The view would then form spaces intermingled with a complex order, or even without order. For this observer the identity of spaces of views based on the observation of the appearances of various groups of objects, could only be a fortunate simplicity in nature.

Summary and Conclusions

Starting from the general idea of a space as a totality satisfying a geometry[1], and having made an inventory of the relations and elementary terms which nature presents, we have begun to investigate the ways in which these relations and terms form spaces, either directly, or by combination.

We have observed four distinct geometric networks in nature. They all naturally have sense data as ultimate terms. Two of these networks have as elementary relations *global resemblance* and *succession*: they concern the pure and simple recurrences of the contents of perception according to the order of sensible time. The first orders the data of any external sense; it merely illustrates, it is true, the rudimentary geometry of *analysis situs*. The second orders the sensations of movement and attitude conjoined with the recurrences of an external datum having any invariable quality whatever: it forms a complete geometric space, or even a whole collection of such spaces.

The last two networks are fabricated from new relations and bring us back to the order of nothing beyond the data of an external sense. But this sense must allow displacement of its data in its field: in other words, the global resemblance of two data must be divided into local and qualitative resemblances, these two partial resemblances diverging and intersecting.

Apprehending the total perception of such a sense in an instant, we have in the first place admitted that the various *groups* formed by the data it comprises present new simple resemblances. Thus, the triads of data would be divided into two classes of aligned and non-aligned triads; or again, each pair would partition the other pairs into two classes according to the presence or absence of a certain kind of equality between them and it. We have united the resemblances of this third type under the name of *relations of position*: these relations could, themselves, transmit the order of a complete space, or, at

(1) Cf. our article Mathematics (Foundations of), *Encyclopaedia Britannica*, 11th ed. vol. suppl.

least, of part of a space, to the content of each total perception of the sense considered.

But the relations of position depend solely on the sensible positions, and not on the rest of the quality, of the data they connect: they are transmitted by local resemblance from the perception of one instant to the perception of another instant. A variant of the foregoing hypothesis then presents itself. If we are disposed to allow that a place in the field of sensation is something more than a class of data resembling each other locally – that it is, for instance, a quality common to all these data, or, to go further, a simple something directly apprehended in them – it becomes natural to suppose that the relations of position connect sensible places directly and turn the sensible field into an immutable space, which each total perception fills and which alone lends to that perception its order.

The relations of position provide the only simple interpretation geometry admits of in nature: in the other spatial networks, all geometric relations are to be interpreted as combinations of relations. Relations of position also provide the only interpretation of geometry which is apprehended in an instant; for no other sensible space is present in its entirety at the same time. This is why the intuitive relations of positions occupy the first place in the image that we form of a sensible structure illustrating a geometry. However, we have observed that it is conceivable that they do not exist at all and are but a dream of the imagination: furthermore, they could only transform a sensible field into a space in the case of a three-dimensional sense that knew no limits: under any other hypothesis, their geometry becomes incomplete. In any case, even if an intuitive space of this type does exist, no matter what sway it may have over the imagination, we have convinced ourselves that inductive sensible spaces comprising temporal relations do not derive their order from it, and that if it can help us to think about them, it cannot in any way manufacture them, even for our intellect.

Next, leaving the relations of position on one side, we have investigated a final ordered network, the richest and most ingenious. It requires *local* and *qualitative resemblance*. On the other hand, it ignores the order of succession of sense data and only takes into account the presence or absence of *simultaneity*: it requires a more specialized resemblance than the first two constructions, but it is satisfied with a more summary temporal relation. We have analyzed it for a three-dimensional, and later for a two-dimensional, vision: we have demonstrated that in its own way it orders the field of sensation, duplicating the network of relations of position if this network exists, thus re-doing its work, and, moreover, infusing geometry into the order of objects and the order of total perceptions.

Let us draw up a comprehensive table of these various kinds of natural geometries by imagining them assembled in the same experience, which we need only assume to be visual and kinesthetic. Let us return to the visual universe of the last chapter. Confining himself to the three relations of local, qualitative and temporal resemblance, the observer sees that the sensible positions are ordered in the same way as the points on a sphere, that the views in parallel families form so many spaces, and that these spaces are embedded in a total space of 'points of views'. This ordered network only requires one set of six objects. It recurs for each set, and the resulting spaces turn out to be identical.

Let us grant this spectator the intuitive perception of simple relationships among those sensible positions which are related by the above network through alignment or congruence: an intuitive order now simply duplicates the experiential order in this first part of the construction.

But now let us remove the bandage which blinds the observer during displacements. A new science emerges: namely, the science of successions of views. As long as we carried him from one view to another without letting him see anything en route, the views followed each other without rhyme or reason for him: indeed, we recall that all the order extracted up to this point was based on the comparison of the contents of various views, without involving their distribution in sensible time. Now, this distribution in its turn presents certain laws. These laws are none other than those of *analysis situs*, whose sensible form has been investigated in the first chapter of Part Three. They partition the views into classes in such a way that one view of a class *a* cannot follow a view of a class *c* without some view of a class *b* appearing in the interval[1]; and they state the properties of this partitioning.

(1) We are neglecting here the deformation of relativity, which would introduce a new richness of order into this sensible world.

We must point out that for these purposes each view is taken as a whole. If two total perceptions resemble each other globally, they are regarded as the same view: if they differ in the sensible position of some appearance, they are regarded as different views purely and simply: whereas the analysis of these differences was previously the whole science, it no longer plays a part in these new laws which only concern the order of the pure and simple recurrences of contents previously encountered. The preceding network determined the content of the unknown views, but not their times of occurrence: this one, by contrast, gives the temporal relations of the occurrences of known views.

These two orders are therefore independent and their coincidence is a remarkable fact. Our spectator could summarize it by observing that the views *in the interval of time* between a view A and a view B derive their *contents* from a class of 'points of views' possessing the formal characteristics of a continuous line from A to B.

Finally we shall stop moving the observer from place to place. Let us allow him to be self-propelled and grant him kinesthetic sensations under the conditions of Chapter II of Part Three. With any view whatever assuming the role of external reference datum, the world of these new sensations is assembled in its turn and in its own way into a multitude of spaces. Furthermore, each view taken as reference gives rise to an analogous and independent network; and all these networks coincide.

This new order in the recurrence of views among successions of kinesthetic sensations has itself no necessary connection with the non-temporal order which the views derive from their contents, any more than with their order of succession. Indeed the coincidence of the spaces of sensations of movement assuming different views as references is reduced to the fact that the same sensation of movement, starting from the same view, constantly leads to the same view. It does not then follow that the pairs of views linked by the same sensation of movement must turn out to be congruent in the space of 'points of views'. Likewise, the fact that the views A and C are always separated in sensible time by some view of a class *b* yields no information about the relation among the sensations of movement connecting these various views, if we confine ourselves to the hypotheses which have proved sufficient for the construction of a kinesthetic geometry: for we have only granted our subject the perception of total impressions of motion, without according him the power of recognizing within the uninterrupted movement leading from a view A to a view C by passing through a view B without stopping, the movements which would lead from A to B and from B to C. But two views separated by the same sensation of movement are separated by the same distance in the geometry of views, and, on the other hand, the views succeeding each other in sensible time are separated by sensations of movement forming a continuous line in kinesthetic geometry: these are *de facto* harmonies among the orders whose mutual independence has been demonstrated and whose various mechanisms have been analyzed.

Such is the geometric structure of one of the simplest imaginable worlds.

This world is still almost incapable of being compared with ours. It is an image of the latter which is simplified by distance, a dream which manifests only a few features. Indeed, what does it not disregard? We have overlooked the imprecision of our senses, the narrowness of their field, the distorting effects of the media. We have excluded the existence of indiscernible objects. We have suppressed any change profounder than a change of point of views. No perceived object is modified or displaced, the environment remains invariable, the sensory mechanism of the observer does not change: there are, moreover, no events apart from apparent events, and no time apart from sensible time.

True, this type of limitation has something naive and inevitable about it. If we wish to form some idea of the order of the data of sense, the intellect immediately agrees on conditions of this type. Select any passage furnishing a sensible expression of a statement of geometry: we shall find that it presupposes the exclusion of all sorts of awkward circumstances, perhaps more tacitly, but not less boldly than we do.

This simplification of the problem is an entirely legitimate method of approach. But we must not forget that the solution obtained by such an abridgement can only be a particular solution. It would be ridiculous to imagine that nature manifests a geometric structure only in the degree to which it conforms to an infantile ideal, that the spatial order of experience concerns fixed objects of distinct and invariable aspect, perceived 'normally' without disturbing influences with the exclusion of moving and changing objects, of indiscernible objects, and of perceptions subject to the influence of heterogeneous or moving media.

This could be true, no doubt, but in fact it is not. It could be that certain objects and certain perceptions of objects are not incorporated in the geometric network of nature which we call space, but this is not the case, and it is the whole realm of the sensible that is ordered by this network. Two objects of identical aspect, an object which is altered and displaced, a view distorted by unequal refraction, a motion diverted by a current, are not outside space.

This amounts to saying that geometry does not apply to the sensible world in a restricted domain of physics, such as, for example, the displacement of invariable solids, but rather in the whole of physics and in each of its branches. Does not every statement of physics involve positions, directions and distances? The geometric network of the world is the network of all its laws, envisaged in certain formal features.

The application of geometry to nature, therefore, does not admit the restrictions we have allowed ourselves. We have shown how certain bodies perceived in a certain way would present a sensible spatial order. But in reality this order encompasses all bodies and perceptions. Let us refrain from saying: Geometry applies to the world only in the degree to which the particular hypotheses in the solutions which have just been presented are satisfied. On the contrary, let us say: These solutions are so far nothing more than particular solutions of a general problem which is none other than that of the empirical import of the whole of physics. They only have value as indications and exercises. They are constructed, not to constrain the intellect, but to enlighten it in the formation of more complete solutions.

In this work, we have hardly begun to take the road which leads us down from schematism to reality. We have been satisfied with positing a primary image of the world: for we must start from something. Three times, however, we have advanced by one degree, abandoning one of the simplifications at first accepted, and coming that much closer to reality. The initial ordered network was then found to be demolished, but in its place appeared a more general network which reproduced its structure without requiring the same restrictions. Indeed, we have passed from a three-dimensional view of a space full of matter to a two-dimensional view of thinly scattered matter: we have observed the obliteration of the initial order, and its resurrection in a more complex form, as well as the emergence of certain new characteristics. Prior to this, for the admittedly rudimentary order of the succession of external data, we have lifted the restriction precluding the existence of identical realities in different places; the network of laws ordering the 'notes' has been broken by exceptions, only to make place for a network of the same form involving 'unities' – much less simple concepts of which 'notes' were particular cases. Finally, in Part One we have outlined the way in which the data of volumes can replace the fictitious punctual data of sensible geometries.

Of course, such reconstructions are easy in comparison with those necessitated by the existence of heterogeneous media, and above all by the radical flux of all things, of objects and organs alike. However, they already provide some idea of what has to be done: they show that the task is arduous, but also that it is feasible.

In short, our hypotheses have hardly transcended the most rudimentary notion of a geometrically ordered sensible world. However, we have investigated it in a systematic spirit with rigour and logic. On the one hand, our original conception of the sensible content of geometry is thereby shielded from those accidental impressions which can give rise to false generalizations. It was in this way that Henri Poincaré, having perceived the existence of an ordered geometric network containing sensations of movement, and not reflecting that a more careful analysis can provide a purely visual manifestation of this same order, concluded that the relation of geometry to experience involves action particularly: a view whose very plausibility results from this hiatus in his examination. Coming from a mind like his, such a delusion shows that one cannot be too meticulous in these initial investigations. On the other hand, precisely because they are no more than schematic, it is advisable that they be exact. Because they will have to undergo considerable corrections in order to apply to nature as we know it, it is essential that they be precise; for we cannot correct what is imprecise. That is why we have presented these initial images of a sensible order, images more difficult to handle than is customary, and yet infinitely simple: perhaps they can already serve to draw an accurate outline, and make any error less plausible.

THE LOGICAL PROBLEM OF INDUCTION

TO MY MASTER

MONSIEUR ANDRÉ LALANDE

Professor at the Sorbonne
Member of the Institute

IN TESTIMONY OF GRATITUDE AND RESPECT

Contents

Preface

by BERTRAND RUSSELL

It is a pleasure, though a melancholy one, to have this opportunity of introducing a reprint of Jean Nicod's work on *The Logical Problems of Induction.* His early death in 1924 was a cause of deep private sorrow to his friends and of public misfortune to the philosophy of science. Although it is thirty-seven years since his death, I still cannot think of him without a sharp pang of grief. He was one of the most lovable human beings I have ever known. But, to the world, the public loss is more important. In his work on induction, as in everything that he wrote, there is a very rare combination of clarity with imagination. These two merits are too seldom found together. Clarity comes more readily in expounding theories which have become conventional than it does in expounding what is novel. Novel theories, more often than not, rise out of a mental mist and only gradually acquire sharp outlines. Nicod's writing, on the contrary, however original, has an exquisite definiteness from which the inquiring reader derives intense aesthetic delight.

The problem of induction which is dealt with in the following pages has been for centuries a scandal in the theory of knowledge and in scientific philosophy. An odd circumstance is that, while it has worried philosophers, men of science, for the most part, have remained completely indifferent to it. Francis Bacon, as a philosopher, gave prominence to the problem, and most subsequent philosophers have respected him for doing so. But when it came to science, Bacon was wrong on almost every point. The great discoveries of his contemporaries were almost all rejected by him – even the circulation of the blood, discovered by his own physician – and certainly were not made in accordance with his precepts for inductive reasoning. The same thing is true in our own day. Philosophers are still wrangling about induction, but scientists will have none of it. At one time, I had the opportunity of frequent discussions with Einstein. He would not concede that anything at all analogous to induction was involved in establishing the theory of relativity. I never quite discovered what, according to him, was involved, but I suspect that it was something more like a sudden flash of illumination. And so the problem remains almost unchanged since the time of Hume. Ever since his time, it has seemed clear to candid investigators of the philosophy of science both that there is no reason to think inductive inferences valid and that, without them, sciences must collapse. Various subterfuges have been used to afford a pretended escape from this dilemma. They continue down to the present day.

Nicod will have none of this obfuscation. He has the courage to arrive at a purely negative conclusion. His work consists in the main in an examination of two proposed justifications of induction. One is what is called the method of elimination, where it is assumed that any definite class of phenomena must be due to some definite class of causes. Sometimes, when the method of elimination is used, the assumption involved is more sophisticated than this, but always, as Nicod demonstrates (conclusively, in my opinion), the desired justification of induction is invalidated by fatal flaws.

The second part of Nicod's treatise, which is the more interesting, is concerned with attempts to justify induction by simple enumeration, more particularly with the very valiant attempt made by Keynes in his *Theory of Probability*. Keynes professed to have found a justification for induction by simple enumeration provided that there was some finite probability in favour of two postulates which he considered plausible. Nicod discovers mistakes in Keynes's use and interpretation of these two postulates and leaves the baffled reader no better off than he was before studying Keynes.

Nicod's treatise is short and by no means exhausts the subject of induction. It cannot be doubted that, if he had lived, he would have discussed aspects of the question which are only hinted at in the work that he was able to do. For my part, I cannot find anything to criticize adversely in what he has written, but I deeply regret that he was not able to investigate such questions as the following: Is induction a law of logic or a law of nature? Is induction valid always or only with suitable limitations, and, in the latter case, what should its limitations be? Should we, perhaps, adopt the somewhat despairing theory of Professor Popper that supposed scientific laws can be disproved, but never proved or even rendered probable?

Does induction require for its validity something which may be vaguely called the reign of law, and, if so, can the reign of law be so formulated as to be not a tautology? All these questions, no doubt, Nicod would have dealt with if he had lived. He would have dealt with them, I am sure, incisively and conclusively.

Finally, I am glad that he is being remembered and that his brief activity is being duly honoured.

BERTRAND RUSSELL

1961

Preface

by ANDRÉ LALANDE

It is with deep sadness that I am now writing what I should have been glad to say for Jean Nicod on the day of the defence of his thesis concerning a book so full of promise as the present one, in which he was testing the foundations and setting the cornerstone of a structure that he was not to complete.

With him there disappears in full youth an unforgettable figure. His friends speak with profound feeling of his heart and character; his professors have been able to appreciate this free and noble spirit, ardent in the pursuit of ideas, in whose nature were compounded the rarest and the most varied qualities. His was primarily, as one of the friends who knew him best wrote me, a sensibility that vibrated to the things of life and of art with the freshness of a childlike soul. He had an extraordinary faculty of enjoying a line, a colour, a sunbeam playing on the leaves of a tree; his was an imagination of a rare and charming fancy which the slightest stimulus aroused to delightful expression. But he retained all these qualities without ostentatious exuberance; if he laughed frankly at what amused him, he himself always spoke with tranquillity; enthusiasm or humour were only accompanied then by an instantaneous light in his eye, or by a hardly perceptible smile around the corners of his lips. He was endowed with an unremitting intellectual curiosity, aided by a facility of understanding and learning such as I do not recall ever having met before. I do not mean that banal ease which comes from mere memory and passive acceptance of the views of others, but rather, that quick grasp which comes from a solid apprehension of ideas, and which serves only as an instrument in the service of the higher power of personal judgement and creative reflection. The rigour of his reasoning was as great as the breadth of his imagination; there was something like genius in the rapidity and scope of his intellect, his friends used to say. The unanimous consensus of his comrades and teachers placed him in the front rank of the new philosophical generation.

Born in 1893 of a family of great intellectual culture, he had at first turned towards the sciences, and he had acquired by the age of eighteen, after two years of special mathematical studies, that solid fund of knowledge and technical habits which are obtained only with difficulty in later education. But philosophy appealed to him and absorbed his interests. He came to the Sorbonne, where in three years he obtained his degree, diploma of graduate studies, and his fellowship. He started his studies as a Fellow first in 1914, in the session interrupted by the declaration of war with Germany. Meanwhile, he had pursued graduate course in the Ecole des Hautes-Etudes, and in the Faculty of Sciences; he had learned both Greek and English so well that he was accorded on his diploma-examination a grade of sixteen for an explication of Plato, and he also carried off first prize at Cambridge University in competition with British students.

Too frail in constitution to be drafted, he spent the greater part of the war period at Cambridge, working diligently on the most varied subjects (he even went so far as to learn Persian in a few months of his leisure time), taking the English degrees, studying particularly, under the invaluable direction of Bertrand Russell, problems of logic and logistics which had already awakened his curiosity during his studies at the Sorbonne. In this realm, to which few French mathematicians or philosophers devote themselves today, he brought the resources of a knowledge and ingenuity which promised an eminent successor to the work which was unfortunately interrupted by the tragic death of Louis Couturat – and which Nicod's departure leaves again in suspense. For what was at first with him only a certain lack of physiological resistance did not take long in assuming the form of a too well known illness against which medicine is almost disarmed. On his return from England he married one of his student comrades, Miss Jouanest, who brought him not only the tenderness of a warm devotion, but also the support of an intelligence capable of understanding his. At first, he followed the usual career of young Fellows: he taught philosophy at the lycées of Toulon, Cahors, and Laon; but the fatigue of lecturing made itself felt and he had to give up secondary teaching. With his extraordinary faculty of learning, and as a result of a competitive examination in which law and political economy played the principal part, he acquired a post, in 1921, with the International Bureau of Labour of the League of Nations. He became noticed quickly for the rapidity and accuracy of his work and the clarity of his mind; his linguistic knowledge made him an invaluable interpreter in international meetings. An improvement in health allowed him to come to Paris for some time where he was able to give a course on the history of Greek philosophy, and where he worked at the same time on his theses. But in the winter of 1922–1923, a rest at Leysin became necessary, and after that, in spite of periods of relative improvement in health, he was no longer destined to resume work. He had just returned to his functions at the International Bureau of Labour at Geneva, his doctoral theses were printed and handed in, and he was to defend them soon after at the Sorbonne, when abrupt complications set in; on February 16, 1924, he was removed from the affection of his family and friends.

Outside of an important article in the *Encyclopedia Britannica* (New Volumes, Suppl. 1) entitled *Mathematical Logic and the Foundations of Mathematics*, he had presented in 1916 a very remarkable paper to the *Cambridge Philosophical Society : A reduction in the number of the primitive propositions of Logic* (Proceedings, vol. XIX, and separate extract by the *Cambridge University Press*). He had also published three reviews in the *Revue de Metaphysique*, one on Goblot's *Logique*, in which certain passages already anticipate the present work; another on the *Geometry of the Sensations of Movement*, in which are sketches of his future work on *Geometry in the Sensible World*; lastly, *The philosophical tendencies of Bertrand Russell*, in which it is seen to what degree he had been attracted to this new conception of logic extending far beyond its traditional limits, fused with mathematics, or more exactly (for we are not concerned here with geometry or even the theory of numbers) with the most general forms of order utilized by mathematics. He was content to deduce the conclusion that such a general logic becomes applicable to the most diverse realms, even to those which would lie outside the forms of space and time, or within the category of empirical determinism; it is so widely applicable that the knowledge of what exists in fact no longer implies a radical empiristic theory of *all* knowledge. It is around a special point of this conception that *The Logical Problem of Induction* is developed: granted that we are constantly making inductions, what are the logical principles that our experimental reasoning presupposes? Nothing is more opposed to the method of Lachelier, in his famous work on the same or related subject, than Nicod's own method, as Nicod himself admits. He frankly gives up, from the very beginning, a definitively fixed theory on which to build a philosophical system. All that he wants, is to make science advance one step further on a difficult ground whose pitfalls, he believes, have not been sufficiently heeded by philosophers. He pursues the method recommended so often by Rauh: he 'takes up the line'. He revises and finds doubtful the formulas which have been current until now; he analyzes and discusses the most recent work which has approached the problem in a really technical way: *A Treatise on Probability*, published in 1921 by John Maynard Keynes, the third part of which is devoted to the relations of probability with induction and analogy. Now accepting the latter's conclusions, then refuting them, but always scrupulously referring to them he tries to show:

(1) that induction by simple enumeration is a fundamental mode of proof and that all those who have thought they can do without it have done so only by the aid of sophisms;

(2) that this style of reasoning would still retain its value, even if determinism were not postulated in advance;

(3) that it can increase the probability of a hypothesis, even when the new facts observed should do nothing but repeat without variation facts already known;

(4) that induction by the elimination of causes can never exceed mediocre probability;

(5) lastly, that is it not actually demonstrated by any procedure whatsoever, how inductive reasoning can raise the probability of a law to the point of indefinite proximity to a certainty.

This is not the place to investigate what might be opposed or added to these conclusions, which partake naturally, except perhaps the first, of the difficulties involved in the indispensable but quite obscure notions of probability. There is little doubt that if the creative and penetrating mind to whom we owe this analysis had been conserved a longer time, he would have pursued the study of this question which is central in the logic of the sciences; and without doubt, the critical part of his investigation would have resulted in a more positive construction. But it was a great step forward to separate distinct problems, to seek the limits and conditions of each, and to show how illusory is the ease with which they were thought to be solved. For such a task nothing less was needed than the fine intelligence whose premature disappearance leaves those who have known him in such deep regret.

ANDRÉ LALANDE

Introduction

Logic is the study of proofs. As the best proofs are to be found in the sciences, it is natural that the logician should keep close to the scientist and fasten upon his reasonings and methods. But the description of these proofs is not his whole task; he must also analyze them. The logician, as a result of the work which falls within his province, can, on his side, teach the scientist something. He can discover for him the elements and premisses of the methods of proof in which he places confidence. For the scientist can use them in complete ignorance of the conditions and causes of their strength; he may believe them to be simple, where the logician finds them complex and possessing an unexpected structure. Thus methodology is only one half of logic. Logic may fall into error through insufficient attention to the proofs of the sciences; it is then working on a poor and inferior subject-matter. But it may also fall into error through an insufficiently courageous analysis of these proofs; that is what has happened in the logic of induction.

At a first inspection we see two types of induction. One proceeds by a simple enumeration of instances. It bases itself solely on their number, and does not claim to derive from that anything but a probability, which may be more or less strong. The other, however, proceeds by the analysis of the circumstances. Being sure of itself, it relies entirely on care, and not at all on repetition, and it aims at certainty. Of these two types of induction, only the latter seems to correspond to the practice and even the spirit of science. A single experience, the scientist thinks, provided that every care has been taken, can bring us at a single stroke all the certainty that is attainable; to wish to build anything whatever on repetition is unworthy of the intelligence.

The logician in general accepts this proud thought too lightly. He closes his mind to the possibility that here things are not as they seem, that the appearance of certainty obtained by the analysis of conditions conceals a probability based on repetition pure and simple, that scientific induction breaks down into an aggregate of enumerative inductions, and therefore that induction by enumeration of instances is the one that he has to justify in order to justify science. Dominated by the impression of power produced by scientific induction, he is no longer willing to recognize any other. He says with Bacon: *Inductio per enumerationem simplicem precario concludit et periculo exponitur ab instantia contradictoria*, and turns away to go in search of the theory of a type of induction which would reckon the repetition of instances as worth nothing and would raise itself to certainty by nobler means.

To do that is to forsake the substance in favour of the shadow. For no induction fits the account given by the doctrine built up in this way. The conditions that it takes for granted are never in fact fulfilled; more than that, they are unreal from the start as a matter of sheer logic. By taking up a position in the field of certainty and wishing to ignore the effect of repetition, the theory of induction finally reaches a complete *impasse*.

It would have been surprising if certain doubts had not arisen. But as they have appeared too late on the scene and lack the strength to withstand a prejudice decked out in the glamour of science, these doubts often give rise only to yet another error. It is realized that the theory that has been elaborated lacks application, and that the strength that it claims to confer on induction must be whittled down to some extent. And so it is said that, though certain in theory, it is only probable in practice, and it is thought that enough has been conceded. But in fact, if actual inductions do not fulfil the conditions that would render them certain and which have previously been taken for granted in the theory, it follows that they are not certain, but it does not in the least follow that, despite that, they remain fairly probable, or very probable, or extremely probable; if certainty is not available, it remains to establish probability in all its entirety and to refashion the theory completely.

We propose in this work to confirm this principle and study the consequences of it. We attempt to establish the logical problem of induction in its true field, the field of probability. We seek for the solution of this problem, but admit that we do not reach it. All that we hope for is to make a contribution to its solution, if only by making it apparent to how great an extent the problem remains obscure – obscure to such a degree that no one has yet got as far as even to enunciate, much less prove, principles that are capable of fully justifying induction under the conditions in which it operates.

Mr Keynes's recent work, *A Treatise on Probability* will often enter into the discussion. We shall attack his fundamental conception of the mechanism of induction and the proof of one of the two essential theorems. But the value of the theorem that has been fully and satisfactorily proved will emerge all the better for that. In it we see the most important result that has yet been obtained; and since we shall have to criticise Mr Keynes several times, let us say here that, in our opinion, no author since Mill has advanced the logical theory of induction as much as he has.

CHAPTER I

Preliminary Notions

It will be useful to begin by defining certain notions and certain principles.

Certain inference and probable inference

What is induction? Induction is a species of inference; but we must make it clear that an inference does not need to be certain in order to be legitimate, and rigorous in its own way. We tend at first to think of inference as the perception of a link between the premisses and the conclusion which assures us that the conclusion is true if the premisses are. This link is *implication*, and we shall call an inference based on it a *certain inference*. But there are weaker links than this, which also provide a basis for inference. No name for them, up to the present, has won universal acceptance. Let us call them, following Keynes, *probability – relations*. (*A Treatise on Probability*, London, 1921, ch. 1.) The presence of one of these relations between a set of propositions A and proposition B assures us that, in the absence of any other information, if A is true, B is probable to a degree p. As before, A is a set of premisses, as before B is a conclusion, and as before the perception of such a relation between A and B is an inference. Let us call this second type of inference *probable inference*.

Certain premisses and probable premisses; general definition of inference

These terms 'certain inference' and 'probable inference' are no doubt open to equivocation. But they seem to be the best available, and it will be sufficient if they are explained.

First of all, a probable inference tells us that, in the absence of all other information, the truth of its premisses renders its conclusion probable to a degree p; it thus tells us less than an inference of the kind that we have just called certain does, but the certainty with which it tells us what it does is just as complete. It is not that in one case the conclusion is probably reached, in the other certainly;

probable inference arrives at a probability as certainly as the other type of inference arrives at a certainty.

Up to now we have considered only the case where the premisses are certain. But every inference in which something may be derived from some premisses if they are taken as certain is one in which something may be derived from those same premisses if they are regarded as only probable; this holds whether the inference concerned is in itself certain or probable. We can likewise say that, starting from premisses which have, taken together, a probability p, a certain inference confers on its conclusion that same probability p, and a probable inference, which would confer on its conclusion the probability q if the premisses were certain, confers on it the probability pq.

Thus we can say that certain inference transfers to its conclusion the total amount of certainty or probability possessed by its premisses taken together, and that probable inference transfers part of it. Inference in general will be defined as an operation which communicates to its conclusion either the whole or a part of the certainty or probability of its premisses; it is only in this broad sense that it is permissible to assert that induction is a form of inference. Let us notice that the conclusion of an inference derives only a certainty or probability equal at most to that of the conjunction of its premisses and, therefore, to that of any given one of them, and in particular *the least certain* of them. This obvious truth will be extremely useful to us.

Definition of induction

What sort of inference is induction? It is usually defined by the logical form of its premisses and by that of its conclusion, as a passage from the particular to the universal.

When an induction starts, then, *in addition to premisses which may have any form and content whatever*, and of which so far nothing is known, there are propositions about this member or that of a certain class A of individuals or species, without its being asserted on other grounds that the members concerned are the sum total of the members of the class A (for perfect induction does not interest us here). What induction arrives at, on the other hand, is a proposition about *all the members*, or about *any given member* of that class; we shall return to that distinction in a moment.

Inductions concerning the relations between characters

The extremely general term *proposition* covers, among others, those which deal with those second-order characters which are relations between characters – such as, in the case of a man, the possession of two eyes of different colours, or again a weight of as many kilograms as the number of centimetres by which his height exceeds one metre.

Let A be the character about which an inductive law frames an hypothesis. We often find that the character whose presence is inferred takes the following form: let *b* and *c* be two classes of characters B1, B2, ... Bn, C1, C2, ... Cn – they might be, for example, temperature and density, the 'values' B, C being different temperatures and densities. Finally, let R be a relation in which each B stands to a C. The character which the inductive inference attaches to the character A often consists in a certain relation R between the two characters, one of class *b* and one of class *c*, which accompany the character A; thus, the chemical composition of a body determines, not its temperature or density, but only a certain relation between the two, one of which continues to be a *function* of the other.

These functional laws are extremely complicated and are very widespread. Does the particular form they take make any difference to the mechanism and logical principles of the inductions by which they are established? That is a matter which we should not be prepared to settle out of hand[1]. But at any rate nothing of the sort has become apparent up to the point at which this work ends. So

(1) Thus induction concerning relations between characters raises in a particularly direct way the problem of the connection between *probability* and *simplicity*. As it happens, there are very many relations by which the members of two given series of characters may be ordered in pairs (as there are many curves passing through a given set of points). Every induction in this field assumes the choice of the simplest relations, a choice which is sometimes explained by the psychological reason that the most simple is the most satisfying. But if induction is to be valid, the choice must be justified, and the reason given does not do this; for what guarantee do we have that the formula that is the most convenient formula in relation to what one knows will prove to be the formula that gives us most probability in relation to what we are ignorant of? We there have a logical problem of the greatest difficulty. Perhaps we do not know with sufficient clarity what probability is, nor again in what the simplicity of a formula consists; it may in fact be defined from several points of view. But neither of the two notions, we may be sure, reduces to whatever most panders to the laziness of our minds.

everything that follows can be taken as applying indifferently to induction concerning characters and induction concerning the relations between characters.

Inductions concerning classes of individuals and inductions concerning classes of species

First of all, let us notice that every functional law has the appearance of a law concerning the species of a genus rather than the individuals of a species. Let B1 and C1, B2 and C2 ... Bn and Cn ... be the characters in classes *b* and *c* linked by the relation R. The law which links the relation R between the characters *b* and *c* to the character A also links C1 with AB1, C2 with AB2 ... Cn with ABn ...; it is thus a bundle of laws fixing the value of the characters *c* possessed by the species AB1, AB2, ... ABn ... of the genus A.

But we thought we should also leave open the possibility, in the definition just given, of inductions dealing directly with classes of species, in the way that other inductions deal with classes of individuals. A number – the number two for example – is, in fact, a species – the species of couples; and so every arithmetical induction involving the verification of a formula for different numbers has, for its immediate domain, the various species of a genus, rather than the individuals of a species.

Will the distinction between classes of species and classes of individuals have no bearing on the logical theory of induction? That is very doubtful. But we shall not reach the point where this distinction makes itself felt, and the substance of this study seems to us to apply indifferently to cases where classes of individuals and to cases where classes of species are in question.

Inductive inferences from these *to* all *and from* these *to* any given one

We must distinguish two forms which the conclusion of an induction about A's may take. In spite of John Stuart Mill's inference from particular to particular, these two forms have not been clearly distinguished. The reason is that they coincide with one another in the field of certainty, and only become separate in the field of probability. Thus, if it is certain that *any given* A is B, it is certain that *all* As are B and conversely, and so the two forms seem to differ only verbally. But if we are dealing with a conclusion that is only probable, a difference of degree, and even an independence, becomes apparent. If it is probable that all As are B, it is much more probable still that any given A is B, as only one risk is involved instead of more than one. On the other hand, when it is probable that any given A is B, it may at the same time be improbable or even impossible that all As are B. That is what happens when it is probable or certain that there is, for example, only one A out of a thousand that is not B. We must, therefore, distinguish, in the general study of induction, between conclusions concerning *all* and those concerning *any given one*; the latter are always the more probable, and are sometimes the only ones that are probable at all.

Primary and secondary inductions

We have made few stipulations about the premisses of induction. While the conclusion is about all the members, or about any given member, of a class A, the premisses must include individual propositions having for their subjects some members of A which are not, so far as we know, all the members that there are; but nothing has been said about the other premisses which need to be combined with them. It may be that the power that induction has arises in various ways according to the form that these other premisses take. We must not, therefore, assume in principle that induction is a single thing and can be analyzed in only one way.

Let us suppose, in particular, that an induction has the conclusion of another induction among its premisses. Let us call it in that case a *secondary* induction, and let us call *primary* those inductions none of whose premisses derives the certainty or probability which it is found to have from induction.

It may be that there are general modes of induction which by virtue of the premisses that they require, have application only to secondary inductions. It may be that these modes of inference are those which are the most certain in themselves, that is, they transmit to their conclusions a greater part of the certainty or probability of their premisses, than the modes involved in primary induction do. But it cannot be over-emphasized that this intrinsic superiority is an empty one. In fact, the probability conferred on its conclusion by a piece of reasoning of any sort is at most equal to that possessed by the least probable among its premisses. The probability supplied by any inductive inference whatever cannot, therefore, exceed the highest probability that primary induction is capable of yielding. For that reason, primary induction must be analyzed before any other kind. For not only is it the logical foundation of induction; it already marks the limit of all the assurance obtainable by induction.

Mill's doctrine provides an excellent illustration both of this hierarchy itself and of the failure to recognize it. Mill regards himself as singling out an extremely powerful mode of induction. But it is a secondary mode since it requires as a premiss the law of causality, and that law, according to Mill, is only a conclusion obtained by the primary and inferior method of simple enumeration. If Mill had as his aim not merely the description of scientific inductions, but also their analysis and the discovery of the principles which underlie them, should he not have announced at this point that, as the whole edifice rested on induction by simple enumeration, it was that which in the end required deeper investigation? Instead, we find that he turned away from it. To him it seemed a barbaric and prehistoric form of reasoning. When he needs to use it as a foundation, he assumes that it yields not only a probability, but actually a probability that approximates indefinitely to certainty as instances are multiplied; and that considerable assumption inspires not the slightest doubt or even curiosity in the most famous of inductive logicians. Further, having once taken it as a foundation, he forgets about it. He seems thereafter to consider that induction by simple enumeration lacks power, and to place his trust wholly in scientific induction. He regards the mode of induction which provides a logical basis for his theory as no more than a historical forerunner, and thinks himself free to look down on it in the same way as Bacon had done.

That part of Mill's doctrine has often been criticized. But the criticisms have, strangely, not fastened on its real shortcoming. Mill might well have been reproached for not taking account of that primary mode of induction to which he owes the premiss of the secondary mode which monopolizes his attention and his admiration. But instead of that, he is most commonly blamed for assigning an inductive foundation at all to the law of causality. That is the objection that has been associated with his theory, and in due course has become a classical objection to it. And so the discussion of this particular view of Mill's has not exposed the exact nature of the gap in his system. It is, moreover, simply a gap, and not a contradiction or fallacy.

For that reason, the logicians who have come after him, having rejected this view, have not taken care to avoid another form of the same shortcoming. The other form consists in expecting primary induction, a mode that is scorned and neglected, to provide not, as previously, the law of causality, but some other premise, that is less universal but nonetheless necessary if the mode of induction that one does admit is to have application. Indeed, it is often thought that, to be able to practise scientific induction, it is not enough to adopt the principle that the effect that one is studying has a cause, but that we must also have some idea of the nature of that cause, in order to narrow down at the first stage a multitude of circumstances, observed and unobserved, and restrict oneself to the systematic examination of some of them. The initial idea that limits one's investigation can only be regarded as an analogy, drawn from what are already known to be the causes of effects of the same sort. And so we return, by another route, to Mill's general position, where he elicits a premiss for his scientific induction from the mass of common knowledge previously possessed, which is itself due to a prescientific mode of induction. And, once again like Mill, people are reluctant to see that the priority thus conceded to this primary mode is in the fullest sense a logical one, and irrevocable. We are apt to think of it as merely historical, and no concern of the pure logician. Thus even today, the distinction between primary and secondary modes of induction, although it is a distinction of elementary method, is not in general currency. Is that not a sign that the thinking on this question has not been sufficiently clear?

On the other hand, once one has recognized that distinction, simple as it is, one can no longer ignore it. Every study of the principles of induction and every doctrine concerning them is then governed by the rule which lays down that the probability or certainty attainable by induction in general is founded on and limited by the probability or certainty of primary induction.

Probability and certainty

Probability is different from certainty not only in degree, but also in kind. For certainty is absolute, while probability is relative. If I judge that a proposition is certain – certain, not merely infinitely probable – no new information can subsequently cast doubt on it, provided that I have not made a false judgement of its certainty in the first place. On the other hand, if I have judged that a proposition is probable to a given degree, a new piece of information may render it more or less probable than it was, without my having for that reason made a mistake. The probability of any proposition is thus relative to such and such a body of information, or rather, to use the precise language of Mr Keynes, it is a relation between the proposition and that body(1).

Common sense has never been unaware of this; it holds that the fulfilment of an improbable prediction cannot justify it, and that chance does not establish a reason. But inductive logicians have not always remembered it.

Is not the failure to bear it in mind the source of the suspicion which, since Bacon, has nearly always surrounded induction by simple enumeration? People have not been content with saying that that form of induction, without analysis, yields only a probability. Modest though such a result is, doubt has been cast on its reality. It seems to have nothing firm about it, and to evaporate into absurdities. If induction by simple enumeration were valid, should we not have to believe that, the longer a man lives, the less his chances of dying? (cf. Keynes, *ibid*. ch. XXI). That mode of induction is also traditionally accused, not of arriving at a conclusion which can claim only probability, which would be true, but of reaching a conclusion whose claims are precarious, in other words, not really reaching a conclusion at all.

Now all the paradoxes which this type of induction seems to engender disappear or cease to surprise us as soon as we remember that probability is relative. On the one hand, the falsification by events of a prediction based on numerous favourable instances does

(1) *A Treatise of Probability*, Chap. I. However, the question is perhaps a little more complicated; is there not an intrinsic probability whereby a proposition recommends itself to our mind to a greater or lesser degree independently of any other opinions we may have? This sort of probability would than be as direct and immediate as certainty; it would differ from it only in degree. We see no reason for not admitting it, and we might call it *plausibility* in order to distinguish it from the probability-relation. But only the latter is obtainable by reasoning, and in particular by induction.

not prove that the prediction was not extremely, even infinitely, probable; it shows only that the prediction was not *certain*. On the other hand, if a prediction is, according to the principle of simple enumeration, probable, relative to data consisting solely of favourable instances, it is so no longer relative to data which include further facts which make the prediction impossible, or very improbable. And so the two sources of paradox disappear; for it is no longer surprising that a probable prediction should be subsequently invalidated, nor again that a prediction of which one knows that it is erroneous, should not be rendered probable by arguments which would have that effect in the absence of such knowledge.

The recognition of this principle that probability is not a quality of propositions but a relation between them takes away from probability the appearance it had before of something fleeting and provisional. It makes it as solid a fact as entailment, for example. The propositions which a given set of propositions makes probable to degree p are as fully determined as the propositions which this same set makes certain, even if they are sometimes as difficult to discover.

But the relative character of probability, even if it does provide a solid assurance of its existence against the doubts which a first glance would suggest, introduces a profound difference between probability and certainty and makes it more difficult to compare them. Thus it is commonly said that probability, as it increases, tends towards certainty as its limit. As a rigorous statement, that is not true. If that were so, infinite probability would be the same thing as certainty. That identity is currently accepted, but it is not a strict one; for there is nothing in the process whereby a probability increases, even carried to infinity, which makes the probability any less relative to a given set of information, or any less subject to alteration in the light of some new information; and this relativity separates it from certainty by an infinite gulf. The probability that an unknown number is not 1324 is infinite and we cannot conceive of anything greater. All the same, it differs enormously from certainty. For it is relative to a state of information in which the unknown number may as well have any value one chooses as any other. If it is learned that there is a probability p, however slight, that the true value is less than 10000, the value 1324 immediately regains the finite probability of one ten-thousandth of p; and if we are told that the first three digits are 1, 3 and 2, the probability becomes 1/10. A probability may well be infinite, but it is still not absolute. It is thus in no way identical to certainty. Is it nonetheless equivalent to it? That is a difficult question, as with any probability, even an infinite one, a risk remains; it is still possible that the unknown number is 1324. This risk is no doubt negligible: it is infinitesimal, it is as small as it can be; but it is not reduced to zero since it exists. As for saying that it is infinitely small, it makes no sense to say that. The expression can in fact be applied only to a function, and it then signifies that for every value a, there exists a value of the variable or variables for which the function is smaller than a. Applied to an individual value, such as that of the risk that obtains under specified conditions, the expression 'infinitely small' has no sense. The distinction between infinite probability and certainty is thus an embarrassing notion. In the rest of this work we shall have occasion to speak, in accordance with usage, of a probability as tending towards certainty; it should be made clear that we shall mean by that simply that it tends towards the highest probability that is conceivable.

Restriction of this work to inductions from these *to* all

Inference being the transference to a conclusion of all or a part of the certainty or probability of the premisses, and induction being an inference whose conclusion is one concerning either *all* the members of class A or *any given one* of them, and certain of whose premisses concern such of those members as do not, so far as we know, constitute the sum total of them, we propose to investigate the logical principles of induction, that is, the other premisses which reason can tell us are required.

In this inquiry we shall mainly be in search of the principles used in those inductions which do not logically presuppose any other inductions – those which we just now called primary; for their principles underlie all inductive reasoning, just as their strength sets a limit to it.

This short study will be restricted to inductions with conclusions concerning *all*. That these are more simple than inductions concerning *any given one* may be doubted. But since these are the only ones that have been studied up to now, they appear to be easier of analysis. Nonetheless, in our view, the difficulties they present have not been recognized, and it may be that fundamentally it is not with them that it would be most expedient for our analysis to begin. (cf. Keynes, *ibid.*, p. 259.) If so, we should have to deviate still further from known theories, and proceed in an entirely new spirit. But before adopting such a course, one must first be convinced that no other course remains.

CHAPTER II

An Hypothesis on Two Elementary Relations between a Fact and a Law

Confirmation and invalidation

Let us consider the law-formula *A necessitates B*. How can an individual proposition, in short a fact, have a bearing on its probability? If the fact consists in the presence of B in a case of A, it is favourable to the law *A necessitates B*; if, on the other hand, it consists in the absence of B in a case of A, it is unfavourable to that law. We may form the notion that it is only in these two ways that a fact can directly affect the probability of a law. A fact which is an instance of the antecedent, either instantiates also the consequent, and gives support to the law, or else it does not instantiate the consequent and denies support to it; these would be the final arbiters in the inductive process. A fact which consists in anything other than the presence or absence of B in a case of A could not, on this view, affect the probability of the law A *necessitates* B *directly*. Instead, since it would consist in the presence or absence of N in the case of M, it would affect the probability of the law M *necessitates* N, and either strengthen or weaken it: it is this last result which would have a bearing, in its turn, on the probability of A *necessitates* B, in virtue of a relation between the probabilities of the two laws, in the case where such a relation, favourable or otherwise, is asserted by some premiss. And so, any effect which individual truths or facts can have on universal propositions or laws takes place through these elementary operations, which we shall call *confirmation* and *invalidation*.

This hypothesis cannot claim the self-evidence of an axiom, but it is one that naturally occurs to us, it makes for great simplification, and reason welcomes it, without insisting on it. We have not seen it enunciated in an explicit way. But we do not think that anything that has ever been written on induction is not compatible with it(1). We can adopt it as a guide.

(1) *Induction by concomitant variations* might be thought of as not being induction by confirmation or invalidation. But such a view could not be upheld. In fact, what is so called consists in an ordinary induction that renders certain or probable the law 'A *variation in A*

The theoretical advantage of invalidation over confirmation

The confirmation that is afforded to a law by a favourable instance is not of the same degree as the invalidation afforded by an unfavourable instance. Confirmation supplies only a probability, whereas invalidation creates a certainty. Confirmation is favourable only, while invalidation is fatal.

Of these two elementary operations of facts upon laws, the negative is thus the only certain one. For that very reason, it is also the more clear and the more precise. Indeed, confirmation by a favourable instance presents two difficulties to the mind which do not exist in the case of invalidation by a counter-instance. In the first place, there is doubt about the very existence of this confirmation when the case which is supposed to supply a basis for it reproduces in every detail a case already used; for it is a common opinion that two verifications that are identical in every respect count only as one. In the second place, we wonder what the measure of this confirmation is when it exists, and do not know what answer to give. Thus the corroborative operation of a favourable instance appears to be enveloped in a certain fog, whereas the operation of a counter-instance seems as limpid and intelligible as it is fatal.

For this reason, a love of clarity and certainty causes our minds to lean involuntarily towards a theory of induction which would rest solely on the invalidating action of experience. An induction can reach a conclusion of certainty only on condition that the elementary operations used are solely ones of invalidation. Where laws are concerned experience is supreme only to establish a denial, and it can attain to equal assurance where affirmation is concerned only where denial involves affirmation. In the second place, this negative action of facts on the probability of laws is the only one which the mind takes in completely right from the start. To make it the sole basis is thus to retain the hope of conceiving of an induction as demonstrative, and also to satisfy reason.

This tendency of the mind seems to us to be linked with two opinions which are more or less universal. According to one, induction must be certain in principle if it is to be probable in practice.

necessitates *a variation in B*', and a deductive passage from that law to '*The reduction to zero of B* necessitates *the reduction to zero of A*', i.e. *A* necessitates *B*, with the aid of a supposed rational principle (which, however, no mathematician could take seriously).

According to the other, the favourable cases or verifications of a law do not corroborate it by reason of their number, but solely by their variety, that being the only thing that reason will listen to. Now if induction is to be certain in principle, it must rest on a negative mechanism. And if it is only the variety of the favourable cases that has any effect, and not their mere multitude, is it not that these cases themselves corroborate only through a process of exclusion? And so the confirmation which instances of a law seem to afford it directly would itself be indirect and essentially negative. The scope of induction would reduce to the invalidation of possible laws by counter-instances.

Almost everything that has been written on induction seems to be in this spirit. Sometimes this principle is avowed, often it is tacitly assumed, but always it imparts a direction to the thinking, and it is undeniable that reason is in favour of it. So let us assume it explicitly and see where it leads.

CHAPTER III

Induction by Invalidation

Mechanism of induction by invalidation : elimination

What is needed is that, after using facts only to invalidate laws, a certain law be confirmed. It is therefore necessary that a number of possible laws be connected in such a way that the rejection of some of them counts in favour of those that remain. In logic this mechanism bears the name *elimination*.

Elimination may, however, be *complete* or *partial*.

If one at least of a group of propositions is true, elimination is complete when all the propositions except *one* is invalidated. The one that remains is then certain, without there being any need to know the initial probabilities of the propositions involved, nor the manner in which the rejection of the first, the second, the third, ... has increased the probability of each of the remaining ones, until finally the rejection of the last but one makes the last certain; the final result does not depend on such considerations.

Elimination is partial, on the other hand, so long as there remain *several* propositions not invalidated. In order to know the value which the probability of one of them has reached, it is in this case necessary to know the initial probabilities, and it must further be assumed that the relative probabilities of the propositions not invalidated remain the same as they were at the beginning. If that assumption is made, the initial probability of the invalidated propositions is distributed among those that remain in proportion to their initial probabilities.

Induction by invalidation can therefore only be used if the conditions for elimination are present. The first of these conditions is a premiss asserting the truth of at least one of a certain group of possible laws. If the facts supply a complete elimination within this group, this condition is sufficient. But if elimination remains a partial one, it is indispensable, in order to evaluate the gain in probability that one of the remaining possible laws has derived from the operation, to know the relative probabilities of these laws.

The deterministic assumption for a given character

Let us begin by discussing the general premiss of every induction by elimination. How can a set of possible laws be constructed one at least of which must be true? There appears to be only one way – by a deterministic assumption. Let A be a character. Let us suppose that *every case of* A *is a case of some other character* X *every case of which is a case of* A or, more briefly, that the character A does not occur without a cause. (By cause, we understand any sufficient condition. By character, we understand any property, which may as well be a relation as an attribute, and may consist in the presence of an antecedent or consequent of a certain sort. On the other hand it would not be sufficient merely to assume that there is some character X every case of which is a case of A, in other words that there is some cause that produces A; for A could then occur as well without a cause, and so we should not be sure of finding some cause of A wherever it occurred. So the required assumption is the above-mentioned one, as we shall see.) To express this assumption, let us say that the character A is *determined*.

Now every case of the character A supplies a group of possible laws of which at least one is true. For if we designate by α the class of characters other than A of the case under consideration, there must be at least one character in class α which necessitates A. The first condition for the use of induction by elimination to support a law of the form X *necessitates* A is thus fulfilled if we suppose the character A to be determined.

Moreover, this supposition can be made narrower still. We can delimit the nature of the characters which potentially necessitate A. Thus one can assume, where A is a character belonging to events, that it is determined in each of its occurrences by circumstances lying in the past, and can assume further that they lie in the immediate past, and even that they lie in the immediate past and occur in the immediately adjacent region of space. That is no doubt the most that we could consider postulating *a priori* – that is, in a primary induction. But in inductions that relate to effects of an already familiar type, much more is commonly taken for granted. Nonetheless, since these further restrictions are based on analogy, they cannot be certain.

Induction by elimination requires a deterministic assumption

The determinedness of the character in the consequent is an essential premiss of any induction by elimination which is used to support a law correlating two characters. This determinedness constitutes the very nerve of the reasoning, the lever by which the rejection of certain laws counts in favour of those that remain. That is a conclusion that will hardly be disputed.[1]

Scope of this assumption

However, three remarks are in place.

Although the majority of authors are ready to admit that induction in general rests on a deterministic principle, we have only established it so far as concerns induction by elimination, whose sphere consists in the invalidation of laws by counter-instances. It may be that the same is true of all induction. But what we have just said can in no way prejudge that.

Secondly, what is of interest for induction by elimination, which has for its object the establishment of a law of the form X *necessitates* A, is not universal determinism, but solely the determinedness of the character A. For, if A is determined, this fact provides us with all that is required for an eliminative investigation, by supplying us, for every case of A, with a class of characters one or other of which necessitates A; it would be no hindrance to the establishment of the law X *necessitates* A by the elimination of the remainder of a group of possible laws if A were the only character in the world that was determined.

Finally, if the determinedness of the character A is the principle in virtue of which the invalidation of the law Y *necessitates* A counts in favour of the law X *necessitates* A, where Y and X are two characters observed in the same case of A – if, in other words, the determinedness of A is the nerve of the process of establishing by elimination a law concerning the production of A – that should not be taken to

(1) However, Mr Keynes seems to be among the authors who do not accept this dependence. Mr Keynes actually thinks that he can show that induction by accumulation of instances can confer on a law a higher probability than the initial probability that it is determined, and therefore its determinedness is not a premiss of such an induction. But Mr Keynes maintains that such an induction has elimination as its principle; he does not recognize therefore, that elimination carries a deterministic postulate and hence cannot confer a probability exceeding that of this premiss.

mean that this operation requires, if it is to be valid, that the deterministic assumption be *certain*.

For if any piece of reasoning confers on its conclusion the degree r of probability or certainty, if premiss A be assumed as certain, that same reasoning will still confer on it a degree of probability r' which is weaker, but not reduced to zero, when the same premiss is assumed to be probable only to a degree s. Any argument which is favourable to its conclusion when its premiss is certain is still favourable to it, though with less force, when that premiss is only probable. That is an indisputable axiom.

In fact, if it is not, as before, certain, but only probable to a degree s, that at least one of the propositions of a given class is true, the elimination still operates as it did before; at every step it gathers together the initial capital of certainty or probability and reallocates it among the alternatives that remain, until this capital finally devolves in its entirety on the last.

Let us suppose then that it is not, as before, certain, but only probable to a degree s, that the character A does not occur without a cause (or without a cause of a certain sort) and let α be the class of characters (or characters of this sort) which accompany A in a given individual case. The observation of one of these characters in the absence of A increases still further the probability that any particular one of those remaining necessitates A; and the elimination of all except one renders the probability that the last, X, necessitates A equal to s. For at this point it is possible to say: either character A occurs in the given case without a cause (or without a cause of the sort assumed) or else X causes A; so there is a probability s that the first limb of this disjunction is false and therefore that the second is true.

It is thus inaccurate to say that the certainty of the deterministic assumption in the case of the character whose cause is being investigated is necessary for the validation of induction by elimination. The probability of this assumption is enough. However slight this probability, induction by elimination has some force, and does render more probable the law in whose favour it operates. The general nature of inference demands this too. For it lays down that the demotion of a premiss from being certain to being probable lessens the force of an argument without destroying it.

But the general nature of inference also sets as a limit to the probability that induction by elimination can confer on a law of the form X *necessitates* A the probability of A's being determined. For an argument can give to its conclusion only a probability at most equal to that of its least certain premiss.(1) It thus remains true that induction by elimination can reach certainty, or indefinitely approach it, only if it is certain or infinitely probable that the character whose cause is under investigation is determined.

It is in that sense, which in any case is simply a matter of conformity with logic, that determinism is a premiss of induction by elimination.

Other conditions of induction by elimination

Let A be a character which is determined (or determined by characters of a certain kind). What more is required for induction by elimination to be used in support of a law of the form X *necessitates* A? It is further necessary that a list be compiled of characters (or characters of that kind) which accompany A in a given individual case. Lastly, the facts must eliminate all the characters in the list except the character X. The law X *necessitates* A is rendered certain when these three premisses are certain, and therefore, when one or other of these premisses is only probable, the law is rendered probable to the same degree as the conjunction of them is. These are the conditions for induction by complete elimination, and such is the power it possesses. It will be seen that an induction of that sort is a certain inference in the sense defined at the beginning.

When the second or third condition is not fulfilled, there remain alongside X other characters which have not been eliminated, and the elimination remains partial or incomplete. If we adopt the general hypothesis that after each elimination of a character, the relative chances of those remaining are the same, the probability conferred on the law X *necessitates* A by the elimination of only some of the competing laws *depends on the initial probabilities* of that law and those that remain. So long as we are ignorant of these probabilities, we are ignorant also of the probability conferred on the law X *necessitates* A by partial elimination; any conjecture we make about

(1) Nonetheless, the probability to be considered is not the probability that A has a cause *in all its occurrences* but the probability that A has a cause in any one of its occurrences; what matters is the probability of the presence of a cause in the individual occurrence of A which supplies the list of possible causes which the elimination works on.

it is a conjecture about the others also. The use of partial elimination does not require knowledge of all the characters which are capable of being the cause of A in a given individual case. But it does require a knowledge of the chances possessed initially by the characters under consideration; and that is a condition that we must not lose sight of.

We now know the conditions for induction by elimination, whether complete or partial. Let us see if the universe fulfils them.

A. INSTANCES OF WHICH OUR KNOWLEDGE IS COMPLETE

Let us begin with induction by complete elimination. Let us suppose that we have a character A which is determined and the list of characters which accompany A in a given case and are capable of being its cause. It may appear that the elimination of all these characters save one is under these conditions only a question of skill or luck. But in fact the operation can still be blocked by a sheer logical impossibility.

Superabundance of causes

Let us suppose that among the characters present along with A in the given case and potentially its cause there is *more than one* which necessitates A; this supposition is possible, since the postulate that A is determined tells us only that there is *at least one* of these characters which necessitates A. Let X and Z be two of them, and let each of them necessitate A. It is then impossible to establish by complete elimination *either of the two laws.* That is obvious, since, both laws being true, neither can be invalidated; but as their truth is not known, induction by elimination comes to a halt with the incomplete result that one at least of the laws must be true, without our being able to say which is, nor whether they are both true.

But is this superabundance of causes in a single instance of an effect a very rare and special occurrence? Far from it; it is the rule and is absent only in two extremely special cases.

The possibility of a complex cause makes complete elimination impossible

It may be that the list of possible causes of A, in the case considered, includes, besides characters L, M, N, some characters composed of several of these, such as LM or LMN; that is what happens every time that the information at our disposal concerning the manner in which A is determined does not with certainty rule out a complex cause.

Let X be the least complex character which necessitates A. Every other character of which X is a factor, such as LX equally necessitates A. Thus once again we come across the superabundance which has just been discussed, and it appears no longer as something exceptional, but as something normal; for it occurs in every case where, in the list of the characters of the instance taken as a basis, the least complex actual cause is not at the same time the most complex of the possible causes.

Induction by elimination leaves in the field as possible laws: X *necessitates* A, LX *necessitates* A, MX *necessitates* A, LM ... X *necessitates* A, without getting to the stage of demonstrating the first. It is true that the last is established, since it is implied by each of the earlier ones. But it must be noticed that it already follows from the very premiss that asserts that A is determined. For the character LM ... X, the most complex of all of them, is simply the conjunction of all the possible causes of A in the case considered, and that conjunction, as A is assumed to be determined, cannot fail to necessitate A.

So even when we assume that a character is determined, and take for granted the list of possible causes present in one of its occurrences, the process of establishing, by the method of induction by elimination, a law governing the production of this character runs into an insurmountable theoretical obstacle in the possibility of a complex cause.

Now that possibility certainly exists; many effects possess causes more complex than themselves. For example, the colour of a compound is determined by the various colours of its elements. We are forced to the conclusion that the elimination cannot be completed, except in favour of an aggregate character which does not need its support, since it has been assumed to necessitate the effect from the start. We remember that induction by elimination, if completed, is the only type of induction which reaches a conclusion which is a certainty.

Partial elimination : a principle directed against the complexity of causes

We have still to see what probability an elimination could give us, which the possibility of a complex cause condemns us not to achieve. Let X be the most simple character (in the list) which necessitates A. We may have eliminated all the characters more simple than X or equally simple. But we cannot have eliminated either X itself, of course, or any of the more complex characters which include X as a factor, such as LX, MX, LMX, etc. The laws X *necessitates* A, LX *necessitates* A, etc. remain in the field. We know that the probability is shared out among them according to their relative initial probabilities.

At this point we might consider appealing to an *a priori* principle of simplicity asserting that, for every degree p of complexity, there is a certain chance π that a given effect has a cause of a complexity smaller than or equal to p, and that this chance π, a function of p, grows in proportion as p increases and tends finally towards certainty. Such a principle would be a plausible one, as it only puts into precise language the acceptable view that an infinitely complex cause is infinitely improbable.

Inadequacy of such a principle

It would give X, the simplest possible cause, a certain advantage over LX, LMX, etc. .... All the same, that advantage would remain a finite one. It would in fact be measured by the value of π corresponding to the degree of complexity of the character X, a value which is limited and finite. This principle would therefore not enable induction by elimination to avoid the obstacle which arises from the possible complexity of causes, except in a very imperfect fashion. It would not put induction by partial elimination, the only sort that is still possible, in a position to confer on a given true conclusion, if not the certainty already found to be unattainable, at least a probability which approximates indefinitely to it.

Now that is what we might expect induction to give us. We can readily accept that induction does not give us certainty. We can accept even more readily that in practice, the limited number of facts at our disposal limits the probability obtainable by inductions. But that a wholly theoretical, and therefore essentially insuperable, reason should condemn induction to stop short at a finite probability, is a conclusion to be accepted only as a last resort.

Another principle directed against the plurality of causes

Another *a priori* principle of simplicity may be thought of, which could help induction by elimination to surmount the obstacle presented by the complexity of causes in an infinitely more effective, though indirect, way. This principle would assert as improbable the *plurality* of causes, not, as before, their *complexity*.

Let us suppose, first of all, that this plurality has been excluded. Let us assume not only that the character A does not occur without a cause, but also that it is always the effect of one and the same cause, so that there exists a character X (of a kind which may or may not be specified in more detail) which is inseparable from A. Induction by elimination can then establish with certainty which this character is. It must indeed be one of those which accompany A in any individual case. But we now have the right to eliminate any character which is absent when A is present, as well as any character which is present when A is absent. We can thus get rid of superabundant characters like LX by showing that even if they are sufficient to necessitate A, only X is necessary. To show that A and X are inseparable, all that is needed is two cases of A having only X in common (neglecting those characters which are not of the specified kind, if one has been specified).

But to postulate in advance that a certain character A has a single condition which is at the same time sufficient and necessary is to make an extremely rash assumption. It is one that is attractive to reason, but not to the extent of appearing to be something that reason can be sure of. And all that is plausible and acceptable in the assumption seems to be contained in the following probability principle: *For every number n, there is a probability N that A carries with it a necessary condition which consists of a disjunction of less than n sufficient conditions, and N tends towards certainty as n increases to infinity.*

We must word it in that way in order to avoid the following complication. If we said '. . . that the number of causes of A is less than *n*', which would be the most natural way of expressing it, we should come up against the fact that when X is a cause of A, any character such as LX is also, and so, in order to enunciate what we mean, we must exclude from consideration superabundant causes of this type. Thus, what appears at first to be the simplest form of words would involve, ultimately, a very considerable complication.

However, the principle is still not expressed in a wholly satisfactory way. The formulation given above is all right when we as yet know nothing about the number of causes of the effect A. But let us suppose that we already know that these causes are more than one – that they number *at least m*; such knowledge would result from observation of *m* cases of A, no two of which have anything in common, or anything which is not known to be incapable of causing A. Either this piece of knowledge puts into abeyance the operation of the principle, or else it leaves it in being. In the first case it is impossible to render it infinitely probable that X *necessitates* A when one knows already that Y *necessitates* A, which seems rather absurd. In the second case, the rather doubtful consequence follows that the more distinct causes of a single effect that are known, the less probable it is that there are still more. The reason is that the principle, as enunciated, goes further than we intend if we assume that it continues to have application to effects of which it is already known that they admit of various causes. All that is required is that if *m* is the minimum number of distinct causes that we know that a given effect A must have, it is infinitely improbable that it has a number of them infinitely larger than this. Let us express this in precise terms: *If it is known that the character A does not possess a necessary condition consisting of a disjunction of less than m sufficient conditions, and the probability that A possesses a necessary condition consisting of a disjunction of less than m sufficient conditions is designated by N, then the value of N tends towards unity as n increases to infinity.* That seems to us to be the most easily acceptable principle. When nothing is known about the plurality of A's causes, the minimum *m* assumes the value 1 and the principle reduces to the one mentioned earlier[1]. That assumption permits induction by partial elimination to approach as near as one wishes to certainty, despite the possibility of complex causes; at any rate in theory, that is, assuming access to all the facts that one wants.

(1) Instead of supposing that N is a function of n, we suppose that it is a function of *m* and *n*. For a fixed value of *m*, whatever it is, N lends towards 1 when *n* increases to infinity, but the value of N for one and the same value of *n* can decrease as m increases.

The indefinite increase in probability through the multiplication of instances

Let two cases of A have only X in common. They will not be sufficient, as they were earlier, to show that X is at the same time a necessary and sufficient condition of A. But it is possible to say: either X is a sufficient condition of A or it is not. If it is not, A possesses two distinct causes in the two cases (since these cases have nothing in common apart from A and X). Any necessary and sufficient condition of A must consist of a disjunction of at least two characters. Therefore, X *necessitates* A is as probable as it is probable that A possesses a necessary and sufficient condition consisting of the disjunction of less than two characters, i.e. a single character. That probability is the value of N for $n = 2$. It is no doubt very small, and it is in any case finite.

But the reasoning may be carried further and applied to more than two cases. If 100 cases of A are available any two of which have nothing but X in common, it may as before be shown that, if X does not necessitate A, any necessary and sufficient condition of A consists of a disjunction of at least 100 characters. It follows that X *necessitates* A is as probable as the falsity of this consequence. Its probability is thus now the value of N for $n = 100$. It is, therefore, clear that as the indefinite accumulation of cases of A any two of which have only X in common confers on the law X *necessitates* A a probability equal to the successive values of N when n increases without limit, it renders that law as probable as one wishes. That is a satisfactory result.

It calls for a number of remarks.

In the first place, the type of induction to which it applies is still induction by elimination, or invalidation. The multitude of cases of AX, on which it bases itself, do not serve the function of confirming the law X *necessitates* A as being favourable instances of it, but they do serve as counter-examples for the invalidation of other laws. And it is only from the invalidation of these that the confirmation of X *necessitates* A is derived, in virtue of our *a priori* principle. The observation of n cases of A no pair of which has anything in common among the necessary and sufficient conditions of A which are initially possible, apart from X, eliminates all those which consist in a disjunction of less than n characters, except for those which have

X as a constituent character. But to say that X is one of the characters whose disjunction constitutes a necessary and sufficient condition of A is to say that X is a sufficient condition of A, or that X *necessitates* A. That is the mechanism of this type of induction; its only scope absolutely always lies in the invalidating operation of facts upon law-propositions.

In view of that, it is remarkable that we observe in it the most striking feature of induction by simple enumeration, which, apparently resting as it does on the confirmatory effect of instances, works on the opposite principle. That feature is the role played in the establishment of a law by the multiplication of instances. The numerical factor introduced here is in no way due to the uncertainty that exists in practice; for complete knowledge has been assumed of the characters present in every instance. In spite of that, to render the law X *necessitates* A infinitely probable, an infinite number of cases of XA are required. It may appear that we are making a direct use of their confirmatory effect, whereas in fact all that we avail ourselves of is the invalidation that they provide of competing laws.

Idea of a theory of induction by repetition

That appearance, and the reality behind it, suggest a theory of induction by repetition. Perhaps induction of that kind is never any different from that which has just been described, or at any rate is an analogous mechanism of reasoning. The favourable effect of instances that verify it would not then be the direct influence on a law that it appears to be. It would itself be only a result of the fatal effect which these instances have when brought to bear on laws which they invalidate. This notion, as we have just seen, is the basis of the view that a new instance only strengthens a law on condition of being different from all the old ones. It is necessary in addition, according to the theory mentioned just now, where a complete knowledge is assumed of every instance, that the new instance should be different from every old one in respect of all of its characters, except the two which the law proposes to connect. In this way we find that variety is at work underneath and is the source and condition of the effect of number, a conception of induction by enumeration which is attractive to reason. It is one that will be encountered again in this study.

Summary

We began by assuming that a character A is determined and that we have a complete list of the characters which accompany it in one of its instances and are capable of necessitating it. Unless we can be sure that not more than one of the characters one or other of which necessitates A does so necessitate it – and we very rarely can – or else, conversely, it is necessitated only by the total character which conjoins all the others – fortunately an exceptional state of affairs – the elimination cannot be completed; and so it is impossible to establish with certainty by induction a law of the form X *necessitates* A. Again, incomplete elimination only confers a definite probability on this law with the help of a special principle of probability. If this principle be directed against the complexity of causes, which is the source of the difficulty, the support it gives is inadequate, as it does not permit the law, X *necessitates* A, supposing that to be true, to achieve more than a moderate probability. If, however, this principle is directed against the plurality of causes, it provides indirectly a satisfactory solution. It does make it possible to render the law X *necessitates* A as probable as one wishes, provided that there is available as large a number as one wishes of instances of XA no two of which have anything in common. It thus makes induction by enumeration dependent upon number, but requires variety as well. These are the results of the critique of elimination itself. They presuppose all the time that the characters of every instance of XA, or at any rate those under consideration, are all known, and that it is certain or at least infinitely probable that the character A is determined. These last conditions must now be examined.

B. INSTANCES OF WHICH OUR KNOWLEDGE IS INCOMPLETE

Our knowledge of instances in nature is never complete

Retaining the deterministic assumption, let us ask if the characters of each instance are as fully known in nature, where quite certainly induction has a valid field of application. To ask the question is to answer it. The circumstances of a fact of nature, whether it be mental or physical, are always known only partially. If no restriction is assumed, the circumstances actually embrace the total universe in time and space, a totality which outruns our knowledge to an infinite extent. But let us restrict ourselves to the immediate neighbourhood and immediate past; a deeper reason makes complete knowledge of this more limited group of phenomena no less unattainable. The reason for it is to be found in the very notion of a circumstance. It is sometimes the case that all or part of the cause of a detectable effect is hidden. It becomes detectable only as a result of an experiment, which consists in an experience in the strict sense of the term, or in the application of an instrument such as a microscope, which is still really an experience. That holds good of mental effects as well as physical. A mental phenomenon may have all or part of its cause in a state which only his reaction to a certain test, or the answers to skilful questions will make apparent even to the subject; it is all a question of dispositions and tendencies.

In the field of facts of nature, therefore, the result of an experiment can be an important circumstance. For that reason, in order to be sure that nothing has been omitted, we should need to have applied all possible experiments. To determine the full state of a piece of matter in a complete fashion and with absolute certainty, it would be necessary to test it with all the substances that will ever be discovered and to examine it with all the instruments that will ever be invented. Similarly, to determine the state of a mind with certainty, even for the subject's own knowledge, it would be necessary to apply to him all the tests that the ingenuity of psychologists will ever think of – a wholly impossible task.

It may, no doubt, seem to us that certain experiments are enough to make a physical state or mental datum entirely manifest. But we cannot be sure of it; for it remains possible that these experiments allow certain differences to escape our observation which have an effect which has not yet been noticed. The electrical state of bodies before its first effects had been met with was an example of that. Most important of all, this assurance, be it more or less strong, and more or less presumptuous, that one knows all that needs to be done in order to reveal the circumstances which it is relevant to take account of, is only founded on experience, that is to say on previous inductions.

In fact how do we know, when we investigate the cause of a certain effect, that we have made a full inventory of the circumstances to be taken into consideration when we have noted certain characters and certain results? It can only be by analogy with the causes we already know of other analogous effects.

Observation cannot actually tell us directly that such and such a character has no part in the production of such and such an effect. So long as one is in complete ignorance of what is operative, one is in ignorance also of what is inoperative; and it is only by indicating what characters must be attended to that experience excludes all the rest by passing them over. That is particularly clear in the case of characters which have not been discovered, like the electrical state before it was thought of. If we judge it to be improbable that a character of this yet unknown sort figures in the list of possible causes of an effect under examination, that cannot be by a direct verification of its lack of influence, as it has never been detected and its very existence is unknown. It is, therefore, the analogy with laws already proved or probable, and that analogy alone, that restricts the probable causes of a certain effect to a known part of the characters immediately adjacent in the past and in space.

The force of this analogy between the unknown causes of a new effect and the known causes of similar effects is not in all cases the same. The variation in it has the result that induction is more or less powerful according to the novelty of the effect and the knowledge obtained of effects of the same sort. The analogy is a precise and strong one in the sciences that are already sure of themselves; it is loose and obscure when its only foundation is the mass of common experience. It comes closest to being absent altogether when the phenomena in question lie outside both established science and practical affairs, like the so-called 'psychic' phenomena; it is then that we see the weakness of induction when it takes its first steps, not knowing yet which direction to take.

Thus, induction imparts to itself a sort of spontaneous impulse, which is strong in proportion to the ground already covered. Its power has a snowball effect. The narrowing down of important circumstances may in the end leave only a single type of character which needs to be taken into account in the production of a certain effect. It is sufficient to discover which character of this type is present for us to judge straightaway that it is the cause that we are looking for. It is then that scientific induction says with pride: a single experiment is enough, provided that it is performed by a man who knows how to direct his attention.

Conditions for primary induction

But logic, which is master, must restore our modesty. The type of induction which uses induction to raise itself up is a secondary induction. No secondary induction, however certain it may be in itself, yields a result that is more certain than primary induction can yield; for no reasoning, even if it reaches its conclusion with certainty, renders its conclusion more probable than the most doubtful among its premisses. If the restriction of the circumstances to be considered in the production of a certain effect is the conclusion of an induction – and we have just seen that it is – induction must be capable of valid application without this help. And the probability that it achieves under these conditions marks the definitive limit of its power.

This question is a matter of logic and not of history. It is not a question of singling out, from among the laws which an empirical science contains, the first fruits of induction when still innocent of the help of analogy. The points where primary induction was applied are of little consequence; it is enough if it is clearly seen that everything rests on it. If we are not careful, through reflecting that a particular law can be established with few experiments by scientific induction – as happens in the courts – we may have the illusion that it is the same with all laws. But we should then be forgetting that a particular one among them can be established so easily only with the help of the analogy of all the others. It would be as if one said that a table could hold itself upright without legs because one can remove any one of its legs without its falling.

Accordingly, for a science of nature to be established by induction, it must be able to carry on by means of phenomena, part of which are unknown, without having any assurance that the part in question is a negligible one. Continuing our study of induction by invalidation, let us see how these new conditions of uncertainty alter the power that we found it to have.

Primary induction by elimination is not satisfactory when applied to nature

Let us suppose in the first place that complexity of causes has been excluded. It is then possible, starting with a case of the effect A, to eliminate all the characters belonging to the case that have been detected, except for the character X, since the right to neglect characters more complex than X, of which X is an element, has been assumed. So long as it was conceded that this complete elimination in favour of X had for its scope the total set of characters present with A in the given case and capable of being its cause, the law X *necessitates* A was established with the same degree of probability or certainty as was accorded to the deterministic assumption in the case of A. But it is now conceded that certain characters escape our observation. The complete elimination that we thought we were carrying out has accordingly been applied to an incomplete list, and is in fact only a partial elimination.

We remember that the result of a partial elimination depends on the values of certain initial probabilities. The initial probability of the characters eliminated is divided out, according to our most general hypothesis, among the characters remaining, in proportion to their initial probabilities. The probability that the law X *necessitates* A derives from the elimination of all the competing laws which have not escaped our observation is not measured by the determination that we have shown in a task of elimination beyond our powers. We cannot say, as we should like, that the law has been rendered as probable as it can be once we have done everything that was in our power. No; probability depends on more rigid rules than the one that uses our conscientiousness as a measure. The probability of the law X *necessitates* A derives exclusively from the intrinsic probabilities of this law itself and the others competing with it.

Of these probabilities we have no idea. To know them, we should in fact have to compare the chances that each of the observed characters has of being the cause of A, not only with those of the rest of them, but also with those of other characters which have escaped our observation and of which we know nothing – not even their number. Their probabilities are unknown. They are no doubt moderate, in view of their large number. They are, in any case, finite; the elimination that has been carried out can thus confer on the law X *necessitates* A only an inadequate probability that does not border on certainty.

A resort to repetition

We cannot rest content with this. For, once again, the strength of this primary induction, which cannot rely on any analogy, sets a limit to all the power of induction in application to nature. It must, therefore, be possible to improve on this result. The means is clear: to repeat the experiments. It is commonly agreed, in fact, that in such a state of ignorance, one must be willing to rely upon the number of instances.

But we are not simply looking for what needs to be done; we are trying to find the mechanism of this counsel of commonsense, and the principle involved in it. Commonsense at this point suggests a strengthening of the moderate and uncertain probability that elimination has just achieved by a procedure which has every appearance of being induction by simple enumeration. But we have already come across the same need to accumulate instances of the law that we wish to establish. We remember, though, that that accumulation did not work in a simple manner and in the way that it appeared to work, but indirectly, and, as before, by invalidation. The same might be the case at this point. Underlying the multiplication of instances which is imposed on us for the second time, we may find once again the principle of elimination, which might in this way suffice, contrary to appearance, to render a law of nature infinitely probable. That is the possibility that we have still to examine.

An attempt to base the effect of repetition on a principle of elimination; a preliminary assumption

This possibility requires that we make afresh an assumption which already in our earlier discussion served as a foundation, the assumption by which a plurality of causes is taken as improbable. So long as no assumption is made about that, the elimination of possible causes of A can come about only by adding to a case of A some cases of *not*-A in which certain circumstances recur and are thus eliminated. But the addition of a number of cases of A cannot eliminate anything so long as we deny all weight to the conjecture that A is due to the same cause in every case. With a view to explaining the effect of the multiplication of instances by an eliminative mechanism, it does appear to be necessary to assume, in a general way, some principle directed against the plurality of causes.

However, the principle does not operate in this context in the way it did before. When we were assuming full knowledge of every instance, the probability conferred on a law by a large number of instances of it any two of which had nothing in common, was the probability asserted by the principle itself that the effect did not proceed from different causes in so many cases; if we had been willing to exclude the plurality of causes, instead of taking it as improbable, two instances might have sufficed to demonstrate a law, and accumulation would have had no part to play. But we are now faced with instances where our knowledge is imperfect; that imperfection by itself, *even if all plurality of causes is excluded*, makes necessary an infinite repetition.

Let us suppose then, for the sake of simplicity, that the effect A possesses a single cause in all its occurrences. Putting side by side two cases of A having only X in common *so far as we know* is not enough to give a satisfactory probability to the law X *necessitates* A. For the probability that it derives from these cases is the probability that the unknown parts of them have nothing in common either; and that probability, which is finite, and no doubt moderate, is totally obscure. But the continued accumulation of cases of XA having no known feature in common appears to be capable of gradually raising the probability of the law X *necessitates* A and of rendering it as near to certainty as one wishes. That must be so, or else the natural sciences must give up any hope of even approaching certainty. For when we are faced with instances of which our knowledge is only partial – and that is so with all natural phenomena – only in their number does there remain any hope of making up for imperfection of analysis. Can its operation be explained by a principle of elimination?

The theory of eliminative probability

Mr Keynes answers yes, and claims that it has no other field of operation. If a second case of XA increases the probability of the law X *necessitates* A, it is because it does, to our knowledge, differ from the first, or there is some chance, for all we know, that it may differ from it. This elimination, whether it be guaranteed or only probable, of some character in the initial case that competed with X as a possible cause of A constitutes the whole of the favourable effect that a second case of XA has upon the probability of the law

X *necessitates* A. In the same way, a third, a fourth or an *n*th case of XA makes a difference only because they eliminate, or have some chance of eliminating, a character common to all the previous cases. It is by this tendency to reduce the part that cases have in common that the accumulation of cases of A increases the chance the remaining character X has of being the cause of A; that, on this view, would be the true field of operation of induction by repetition.

Let us see what Mr Keynes says: 'The whole process of strengthening the argument in favour of the generalisation φ *implies f*[1] by the accumulation of further experience appears to me to consist in making the argument approximate as nearly as possible to the conditions of a perfect analogy, i.e. the assembling of a group of instances of φ having nothing in common by steadily reducing the comprehensiveness of those resemblances φ between the instances which our generalisation disregards. Thus the advantage of additional instances, derived from experience, arises not out of their number as such, but out of their tendency to limit and reduce the comprehensiveness of φ_1 ...' (*A Treatise on Probability*, pp. 227–228). And again, 'I hold, the, that our object is always to increase the negative analogy, or, which is the same thing, to diminish the characteristics common to all the examined instances and yet not taken account of by our generalisation. Our method, however, may be one which certainly achieves this object, or it may be one which possibly achieves it. The former of these, which is obviously the more satisfactory, may consist either in increasing our definite knowledge respecting instances examined already, or in finding additional instances respecting which definite knowledge is obtainable. The second of them consists in finding additional instances of the generalisation, about which, however, our definite knowledge may be meagre; such further instances, if our knowledge about them were more complete, would either increase oı leave unchanged the Negative Analogy; in the former case they would strengthen the argument and in the latter case they would not weaken it; and they must, therefore, be allowed some weight.' (*Ibid*., p. 234.)

The theory of induction by repetition which these two passages summarize in a very clear manner is, to be sure, a seductive one. It justifies the opinion, common among philosophers, that several

(1) Designated in the text by g(øf).

instances which we were sure did not differ from one another in any respect would do no more than a single one would. It humours the idea that induction by multiplication of instances does not possess the status of a genuine principle, but is effective only to the degree to which it imitates induction by analysis. It brings all induction under the elementary operation of invalidation, and that doctrine has a certain clearcut character that is pleasing to the mind. The theory that Mr Keynes propounds is thus without any doubt the natural one. We must nonetheless examine it.

Development of the deterministic theory

We showed at the beginning of this study that every induction by invalidation in support of the law X *necessitates* A has as a premiss that the character A is determined, that is, the proposition that any case of A is at the same time an instance of at least one character which necessitates A.

But it is conceivable that we possess knowledge of a more definite kind. Let there be a certain instance of A. We have just assumed that α, *the class of all the characters in this instance* (*other than A*) includes at least one character which necessitates A. It is not inconceivable that we should be in a position to say as much about a *more restricted* class. It is possible that we should know that a class a of characters, which is only a part of class $\varkappa$, taken by itself includes at least one character which necessitates A. It is possible that we should know that about *several* sub-classes, a_1, a_2 ... consisting of characters of the instance of A under consideration. For it may be that A possesses several sufficient conditions in each of its occurrences. Finally, it is possible that about some of these classes $-a_1$ and a_2 for example – we are *certain* they include a character which necessitates A, and about others – a_3 and a_4 for example – the same thing is merely *probable to the degrees* p_1 *and* p_2 *respectively*.

That is what comes about according to the assumption that is commonly made about the determinedness of the characters of a natural phenomenon. Indeed, it would amount to very little if we contented ourselves with assuming that any one of the characters of a phenomenon is necessitated by some other; for the characters of a phenomenon, if no restriction be placed upon them, include its relations to all other phenomena past, present and future. So normally a good deal more is taken for granted.

It is taken as certain that any present character of a phenomenon is necessitated by one at least of the other characters that it has which belong neither to the future nor even to the present, but solely to the *past*; further, that it is necessitated by one at least of the characters which belong solely to such a date, or more exactly, to *such a stretch of past time as one cares to take*, *however short*; for we consider that the state of nature in any period of time, however brief it may be, determines its state at all later times. Again, it is sometimes assumed that every character of a phenomenon is necessitated if we concern ourselves with a segment of the past that is sufficiently close to it, by one at least of the characters which relate to a *fairly restricted region of space* around the phenomenon being studied(1).

We can then represent to ourselves the way in which the characters of a natural phenomenon are determined by the familiar image of the concentric waves produced by a stone thrown into a pond. The only thing is that we must imagine the sequence of events in reverse – the waves running together from the periphery to meet at the place of the disturbance, of which their arrival is itself the cause. So the conditions capable of determining an event occupy at a given earlier date a region which is extended in proportion to the remoteness of the date. Running up, so to say, from depths of the past, they meet around the event and converge towards the very space which it fills.

Within the class α of characters of an instance of A, those which relate to some stretch of the past (further restricted, if we allow the last assumption, to a finite region of space) constitute therefore a sub-class *a* about which it is *certain* that it includes at least one character which necessitates A. There is thus an infinite number of these *a*-classes within the total class α.

That is not all. Let us consider the class of circumstances in the given instance of A which belong to the present, as A itself does – in other words the circumstances which are *contemporaneous* with the

(1) It is clear that this postulate is necessary in order to remove the influence of possible causes that escape our observation on account of their distance; it is expounded very clearly by M. Painlevé in the article *Mécanique* in the review *De la Méthode dans les Sciences*. We may notice that the postulate directed against action at distance is, however, a consequence of the principle according to which all causation is transmitted through adjacent objects.

effect and not earlier than it. Is there not some probability that the circumstance A is necessitated as well by one at least of them?

Let us begin by showing that this probability could not be a certainty, as in the previous case. That much is obvious, and can furthermore be proved.

A character is in fact at the same time a conjunction of characters. Thus all the characters of a phenomenon which are simultaneous with any character M constitute a character as a group, and if it were *certain* that every present circumstance and therefore the one in question, M, is necessitated by some present circumstance it would be certain that any particular one of the present circumstances of a phenomenon necessitated all the others. It would be certain that its recurrence necessitated theirs. Now that is plainly contrary to the facts. In our universe, many characters are found together without being connected. There are, therefore, some characters, complex ones at least, which are not necessitated by any character simultaneous with them. For that reason, it could not be certain *a priori* that any given character is necessitated by some character simultaneous with it.

We can also readily admit that the *a priori* probability of this, though not negligible, is modest and very far from being equivalent in practice to certainty. Let us designate it by Π. Similarly, there is a probability π that any given character necessitates some character simultaneous with it; and that probability π equally is neither very high, nor is it negligible.

Application to induction by elimination

In a general way, if the character A is necessitated (with a certainty or probability p) by at least one member of a sub-class a of characters which accompany it in a given instance, the search for a character necessitating A can be carried out by an elimination operating on the characters that are members of a. But once all the characters except one have been rejected, we know with the degree of certainty or probability p, and with that degree alone, that the remaining one necessitates A.

In consequence, if the class a is the class of circumstances, in the instance adopted as a basis, which relate to a stretch of time earlier than the occurrence of A, the elimination of all these circumstances except one can make it *certain* that the last necessitates A. *Per contra*, if the class *a* is the class of circumstances *contemporaneous* with A in this instance, the elimination of these circumstances except one renders it probable only to the degree Π that the remaining ones necessitates A.

These results will be useful to us in a moment.

Could probability obtained by elimination be the principle of induction by repetition in its application to nature?

It is not without some hesitation that the proofs that follow are presented. They are longer and more complex than might be desired; they require an effort of attention. But, as I believe them to be correct, it is appropriate that I should set them out. We must follow our reasoning wherever it leads; ὅπη ἂν ο λόγος ὥσπερ πνεῦμα φέρῃ, ταύτῃ ιτέον.

If induction's only field of operation is invalidation, if its sole mechanism is the rejection of possible causes in favour of those that remain, in other words, elimination, if the favourable effect of a new instance of the law X *necessitates* A consists wholly in the sure or probable elimination of some fresh member of a class *a* of circumstances present with X and A in the initial instance, of which it is certain or probable to the degree *p* that one member at least necessitates A, the following consequences result.

We have shown already that this view requires some principle directed against the plurality of causes of the character A, where it is a question of rendering probable a law concerning its production. Let us make the maximum assumption; let us exclude all plurality of causes when we assume that there exists at least one character (simple or complex) which necessitates A and conversely is necessitated by it. Let us take it as certain that in any instance of A a subclass *a* of circumstances includes such a character. In accordance with what was said just now, that class cannot be the class of circumstances contemporaneous with the effect A. It must be the class of circumstances belonging to some *earlier* period of time, which will most naturally be thought of as extremely close to the appearance of A itself. We shall call the members of this class (for we shall consider only one) the *antecedents* of A in the instances concerned, and we shall call the characters contemporaneous with A the *concomitants* of A.

We are thus assuming that it is *certain* that one at least of the antecedents of any character is inseparable from it. On the other hand, it remains *probable to the degree* Π that a given character is implied by one at least of its concomitants, and *probable to the degree* π that it necessitates at least one of its concomitants, the values Π and π being neither negligible nor equivalent in practice to certainty. That set of assumptions is without any doubt the maximum that we could consider assuming *a priori* concerning the manner in which the phenomena of nature are determined. If we succeed in showing that the theory under discussion is not satisfactory in the conditions that are most favourable to it, we shall have adequately proved that it is not so in fact.

The probability conferred on the law X *necessitates* A (X being an antecedent of A in a given case) by a number n of instances of it is, on the view now being examined, a probability of the second degree: it is the probability that the elimination of A's antecedents would reach a certain state in the course of n instances of A, a state which would itself give a certain probability to the law X *necessitates* A. It is by this argument that induction by multiplication of instances, instead of being a genuine form of argument, could only be the shadow of one.

This idea needs to be more precise. For we do not know how far our n instances of XA carry the elimination of A's antecedents; perhaps X is the only one that is retained; perhaps another is allowed to remain, or two others, or x others. But these different suppositions may not be equally probable. On each of them, the probability of the law X *necessitates* A is the probability that the supposition would give to it, if the supposition were realised, multiplied by the probability of its being realized. The global probability conferred on the law is thus some mean value of all these products, smaller than the largest of them.

If that probability is to tend towards certainty as n grows, one of these products must do so as well. And for that to occur, the same must hold of each of its constituent probabilities. It is therefore necessary that the multiplication to infinity of instances of XA should render it infinitely probable that elimination will reach a state which itself renders the law X *necessitates* A either certain or infinitely probable.

That state of affairs could be realized only in one of the following two situations. *X is the only antecedent of A which is common to all the instances considered :* X *necessitates* A is then certain. In the second case, some antecedents of A other than X remain which are common to all these instances. But *it is infinitely probable that these other antecedents do not necessitate A, or else they are necessitated by X.* In the one case, as in the other X *necessitates* A is in fact infinitely probable.

We can here neglect the first alternative.(1) For the antecedents in question, other than X, are characters which have escaped the means of observation that we are using. Nothing is known about their nature, and it could not be less probable that A is necessitated by one of these than it is necessitated by X, still less infinitely less probable.

A state of elimination that renders X *necessitates* A certain or infinitely probable is therefore a state in which *X is left alone* in the field, or else *the other antecedents of A that remain are such that it is infinitely probable that they are necessitated by X.*

Does the multiplication to infinity of instances of XA render infinitely probable the realization of one or the other of these situations?

It could not render the realization of the first infinitely probable. In fact it does not diminish at all the initial probability that X necessitates some concomitant that is to be found among the antecedents of A which escape observation; and that probability p, though not a very high one, is certainly not negligible. Given an infinite number of cases of XA which so far as we know have no character in common, a finite probability therefore remains that is *at least* equal to p that all these instances have, unknown to us, one or more other antecedents of A in common, a set we shall designate by Y; for the probability of a state of affairs is at least equal to the probability of any hypothesis that implies it.

But we must be careful to notice that even if the existence of an unobservable concomitant that is necessitated by X implies the presence of a similar unobservable concomitant in all the instances

(1) It is the one that was met with when we were studying the mechanism of repetition for instances of which we have complete knowledge and for the case where a plurality of causes is improbable in proportion to the size of the plurality.

of X under consideration, the converse implication does not hold; it is obviously possible that a character Y should accompany X in all the instances under consideration and yet forsake it in others. It is true that supposition becomes infinitely probable in proportion as the instances considered are multiplied. But let us be on our guard against using at this point the very principle of induction by repetition, whose justification is at issue in the theory we are examining.

Is it then, on this theory, infinitely probable that any character Y that is a concomitant of X in an infinite number of instances is necessitated by X?

That is what the theory of probability by elimination has to prove; that is what it does not prove and cannot prove, since it is true only if it is false. Let us try to make this contradiction plain.

We suppose that X and Y are the sole antecedents of A present in an infinite number of cases of A, and we ask if this supposition renders it infinitely probable that X *necessitates* A.

Now X is not an antecedent, but a concomitant, of Y. It is not *certain* that Y is implied by any one of its concomitants, but *only probable to the degree* Π. And so, *on the theory under discussion*, the supposition that all the concomitants of Y, except X, have been eliminated does not render X *necessitates* Y certain or infinitely probable, but only probable to the degree Π.

We might seek to dodge this conclusion by objecting that the concomitance of X and Y in an infinite number of cases renders it infinitely probable that the *antecedent* which necessitates X necessitates Y also. But we are now involved in a circle. For we have proved that it is not infinitely probable that all these cases have only a single antecedent in common, even if there is only one to be seen. And it is clear that if there is more than one of them, there is a finite probability that one of them necessitates X and another Y, unless it is infinitely probable that they all necessitate one another; we are thus back where we started.

The whole argument might be summarized as follows. On the theory according to which induction by repetition has eliminative probability as its principle, it is not infinitely probable that an antecedent preceding the effect in an infinite number of cases is its cause, because it is not infinitely probable that this antecedent does not necessitate, unknown to us, another simultaneous with it, but nor could it be infinitely probable that, if another antecedent actually accompanies the first in an infinite number of cases, it is then necessitated by it; for it is not infinitely probable *a priori* that a character is necessitated by its concomitants.

That is the chain of reasoning that seems to us to establish that the theory of Mr Keynes and of philosophical commonsense about the mechanism of induction by simple enumeration does not achieve its object. It does not allow that form of induction to confer on laws of nature a probability higher than the *a priori* probability Π that any character, whether simple or complex, is necessitated by some concomitant character; and that probability Π is a long way from certainty.

But we have earlier shown that primary induction, which is the foundation of all our empirical knowledge of nature, could not yield a probability approaching certainty unless it were derived from an infinite repetition. It is thus shown that the idea of basing all induction on a principle of invalidation and elimination, an idea so attractive to our minds that it seems always to have reigned over them without any resistance, leads in the end to a complete impasse. It is an impasse if we can offer physics a probability that is in principle only a modest one, and one that is separated from certainty by an irreducible gap. It would be very pessimistic, to say the least, to attribute such a cruel restriction to the very nature of things, and Mr Keynes had no such intention.

It is interesting, in order to form a better idea of the precise scope of these results, to compare them with the theses maintained by J. Lachelier in his *Fondement de l'Induction.*

The ideas of J. Lachelier

This author seeks to formulate, and then prove, some principles which appear to our minds to be such as will justify induction, without pausing to make clear the precise manner in which those principles are actually to be applied to inductions, so as to confer on them a determinate probability; such an analysis and confirmation are in our opinion fundamental.

He begins by putting forward the classic thesis according to which the essential premiss of induction is the determination of phenomena by their antecedents. But – and here we have his particular thesis – that first principle is not enough; for, among the antecedents necessary for the production of a certain effect there may in fact occur – we know well that there do occur – some which are undetectable. When we assert that the recurrence of such antecedents as are detectable must necessitate the recurrence of the effect, we are plainly supposing, in virtue of another principle, that all the antecedents required are in fact found together in at least the majority of cases. Lachelier illustrates that with the example of the biological law according to which like engenders like; then, after observing that the intervention of imperceptible conditions is to be found no less in physical or chemical phenomena than in the phenomena of life, he shows that his principle of the cohesion in groups of simultaneous characters is necessary, by the same token, for all the inductions in the natural sciences. He thinks of this mutual determination of concomitant circumstances as 'a principle of order which watches over the maintenance of species.' He sees an element of purpose in it, and sums up as follows: 'We could thus say in a word that the possibility of induction rests on the double principle of efficient and final causes'. (*Du Fondement de l'Induction*, p. 12).

He sees perfectly well that his second principle cannot be, as the first was, a principle of certainty, but is only a principle of probability. In fact the cohesion of concomitant characters is limited; only certain groups form 'species' which keep themselves in being. We cannot, therefore, be certain that the recurrence of the characters that are observable by the means available ensures the recurrence of those imperceptible characters which perhaps are necessary to bring about the effect.

Lachelier leaves the matter there. He thinks that he has solved the strictly logical problem of induction, having formulated principles which plainly do justify it. For him, as for most people, the main thing is not to see what the principles of induction are – they think that too easy – but rather to prove these principles; 'They leave the things and run to the causes'.(1) In his haste to pass on to this metaphysical task, he does not notice at all that the principles whose proof he is searching after are in no way adequate to justify induction.

(1) Montaigne III 11. – Thus we find that Lachelier devotes twelve pages to the formulation of the principles of induction, eighty to their proof.

It is not in fact *certain* that the recurrence of the detectable antecedents of an effect will ensure the recurrence of imperceptible antecedents that are necessary if the effect is to be reproduced. It is therefore only probable; and in Lachelier's principle of final causes we can recognize the assumption that lays down that it is probable *a priori* to a degree Π that a given character is necessitated by its concomitants. Thus, all that results from these two principles of Lachelier's is a moderate probability that the recurrence of the antecedents observed in a given case ensures the recurrence of the effect observed in that same case.

That result is not enough. We must be able to improve on it. In fact we can do. How? By multiplying instances. For it is agreed that the initial probability that the observed antecedents necessitate the observed effect – and that is all that we can derive from Lachelier's principles – is capable of being reduced to nil by a counter instance, and, conversely, increased by favourable instances until it approaches certainty.

Lachelier does not stop to consider this latter operation of facts, and that his principles take no account of it. According to his theory, the probability of a connexion between a consequent and an antecedent depends on that of a connexion between the antecedents themselves – concomitants in other words. His second principle supplies a certain *a priori* probability of a connexion between any two concomitants. But what assurance do we have that the concomitants *observed together the largest number of times* have the largest chance of being connected, or that the probability of their being connected can by this means be raised above the *a priori* probability that one or other of them is connected with some concomitant, and tend towards certainty? That question is not answered by him, or even asked. Yet it is the most important one, as it is the most difficult.

Could eliminative probability be the principle of induction by repetition in its application to numbers?

First of all, does induction by multiplication of instances have application to numbers? Does a multiplication of the verifications of a formula of a law that is in itself uncertain confer on it a probability that grows larger and larger and tends towards certainty, in the field of numbers as well as in nature?

The possibility, in this field, of certain demonstrations, and the unique value attached to them mean that induction by instances is not in principle necessary, and is not officially accepted; mathematics having once tasted certainty, is not content with anything less. They also cast doubt on the very validity of induction by instances. It is thought, not only that such conclusions in arithmetic express only a probability, but also that that probability is precarious and lacks substance. We can light upon a suggestion and use it to guide us, but in our research, not in the establishing of a theorem. We can make use of it in discovery – at our own risk and peril. But we cannot rest any degree whatever of assurance on it.

This view, on reflexion, seems irrational. The precariousness of probability founded on instances is in fact no more a real one in arithmetic than it is in nature. It stems, in this case as in the other, from the failure to remember that probability is *relative to the information available*. Thus, the discovery of a demonstration of the truth or falsity of a law of which we knew only numerous verifications cannot invalidate the fact that the state of information in which we were before rendered that law highly probable, and no more than that.

But, most important of all, if instances provide no legitimate foundation for probability, then those mathematicians who follow such indications in their search for theorems – and the most famous of them have not failed to do so – are not behaving reasonably. Mr Keynes makes this point well: 'Generalisations have been suggested nearly as often, perhaps, in the logical and mathematical sciences, as in the physical, by the recognition of particular instances, even where formal proof has been forthcoming subsequently. Yet if the suggestions of analogy have no appreciable probability in the formal sciences, and should be permitted only in the material, it must be unreasonable for us to pursue them. If no finite probability exists that a formula, for which we have good empirical verification, is in fact universally true, Newton was acting fortunately, but not reasonably, when he hit on the Binomial Theorem by methods of empiricism.

'I am inclined to believe, therefore, that, if we trust the promptings of common sense, we have the same kind of ground for trusting analogy in mathematics that we have in physics, and that we ought to be able to apply any justification of the method, which suits the latter case, to the former also.' (*A Treatise on Probability*, pp. 243–244.)

Is this view not a reasonable one? It invites us, and authorizes us, in every case, to examine in relation to numbers the theory maintained by Mr Keynes, on which eliminative probability is the principle of induction by simple enumeration.

The inadequacy of the theory in this field seems fairly clear. We no longer have to take into account temporal distinctions when we consider the manner in which characters are determined.

Is it *certain* or *infinitely probable* that a given general character of the number n, a character consisting of one or more properties, such as that of being first, or perfect, or a square, is necessitated by some other general character of the same number n?

No, it is neither certain nor infinitely probable, but only probable to a finite degree p. For we know that the general characters of numbers fall into several groups and it is not enough to fix one of the characters of a number for all the others to be fixed. It follows that there is a finite and not a negligible probability of the value $1-p$ that a given general character A is not implied by any other and constitutes an independent group if it is complex, or even, if it is simple, a fundamental property that is not derived from any other.

According to the theory under discussion, 'the whole process of strengthening the argument in favour of the generalisation φ *implies* f ... consists in making the argument approximate as nearly as possible to the conditions of a perfect analogy, by steadily reducing the comprehensiveness of those resemblances φ_1 between the instances which our generalisation disregards' (the passage already quoted on p. 222). And by perfect analogy, Mr Keynes understands a set of two or more cases of XA which eliminates all the rest, i.e. they have no other character in common.

Any collection of numbers exhibiting the two properties X and A no doubt always has other general common properties. We must therefore say, if we follow Mr Keynes, that such a collection, however numerous it is, is not a case of perfect analogy, and can only constitute an argument that is inferior to the perfect analogy that is the ideal and the limiting case. That is the thesis that he is defending.

But what probability would even a perfect analogy confer on the law X *necessitates* A? If, *per impossibile*, we knew that two numbers *m* and *n* had no general property in common other than X and A, what probability would result thereby for the law X *necessitates* A? Would certainty result? Would infinite probability? No, it would only be the probability p. There is in fact always a probability $1-p$ that the general property X does not derive from any other, and therefore does not derive from A. All that a perfect analogy can prove is that X is the only property that *could* imply A. But from that X *necessitates* A follows only to the extent that it is probable that A is not an independent group of general properties of numbers or a fundamental general property that no other property necessitates.

The probability that a perfect analogy would establish is, according to Mr Keynes, the ideal limit of the probability that the multiplication of instances can give us. *The latter probability does not, therefore tend towards certainty, but is always lower than the finite probability p* – an unsatisfactory result.

Conclusion of the study of induction by invalidation

Let us sum up the foregoing analysis.

Having made the hypothesis, if it is one, that any effect that facts have upon the probability of laws resolves itself into these two elementary operations of confirmation or invalidation by favourable or unfavourable instances, we outlined the theoretical advantage that there would be in conceiving of invalidation as the sole sphere of competence of every inductive inference. We observed that philosophers, and indeed reason itself, inclined in that direction. We tried to trace that doctrine to its principle and resolutely follow it up, with a view to deciding whether it is ultimately tenable.

We first settled the essential and necessary form that induction by invalidation must take: it is the transference to one of a given group of laws by the rejection of all or some of the others, of all or part of the certainty or probability of the existence of at least one true law in the group. Such groups of laws that certainly or probably include a true law among them are supplied to induction by some deterministic assumptions. It must be assumed that it is certain or probable to a degree p that in any one of the instances of the character A, for the production of which it is proposed to establish a law, there is at least one character which necessitates A within a certain class α, which may include all the characters of the instance or only some of them. Starting, then, from some instance of the character A and the related class α, induction seeks to invalidate by counter-facts the connexion between A and as large a number as possible of the characters in α, thereby transferring the chances they had initially of necessitating A to those that remain. That is the form that *induction by elimination* takes; on the conception we are studying, it is the only form of induction that there is.

We examined the functioning of this form of induction under ideal conditions where there is full knowledge of the individual cases, and none of their circumstances are hidden from us. But it still comes up against the complexity of causes, if that possibility is admitted. It is therefore only through the indirect help of some assumption directed against the plurality of causes that it can approximate to a certain inference. If we venture to exclude that possibility, certainty is once more attainable. But if we make only the more limited assumption that such plurality is improbable in proportion to its size, elimination provides us once again with a form of inference that is infinitely probable, but on condition of resting on an infinite number of different favourable instances.

We then passed on to consider how things are in nature. No instance is there known in all its circumstances, for the essential reason that a circumstance includes not only what is perceived at the present moment, but also what would be perceived as a result of an experiment and in no case can one carry out all the relevant experiments, or rather, it is not possible to be sure that all the relevant experiments have been carried out. Under the conditions of primary induction, where we have to renounce all the help afforded by empirical knowledge, in order to be in a position to provide it with a valid foundation, induction by elimination cannot derive all that it needs from phenomena of which its knowledge is incomplete, and incomplete to a degree which is itself unknown.

It seems, therefore, that the doctrine that finds the sole sphere of competence of induction in the invalidation of laws by counter-examples must at this point be abandoned. The probability approximating to certainty that seems to be attainable by the natural sciences, must in principle be sought in confirmation by favourable instances, that is, in induction by multiplication of instances.

When we turn to that form of induction itself, the doctrine we are concerned with seems to retrieve its position. Here too, everything is reduced to elimination; but an elimination that is only probable, and would be applied, in our ignorance, to the unknown part of the instances. It is only the probability of this covert elimination that would produce the seemingly direct confirmatory power of collections of favourable instances; it is only in virtue of the probability that two cases that are indistinguishable to us do in fact differ, that they count for more than a single case. Reason inclines towards this theory, and Mr Keynes has the merit of having explicitly enunciated it.

We showed first that the help of a principle directed against the plurality of causes is required. We assumed that every character has a single cause, in order to remove that preliminary difficulty and so examine the very principle underlying the theory of eliminative probability.

If we follow this theory, we find that all induction by repetition has as an ideal and a limit a certain form of induction by elimination, to which it remains consistently inferior. It follows that where it is probable *a priori* only to the degree p that a character is determined by some other character, the accumulation of instances can only give a probability lower than p that two characters are connected; for that value is the maximum that elimination itself can give. That is the case that we thought we had found once more in the field of numbers. Where, however, the manner in which an instance is determined is such that there are several sub-classes within the total class of its characters, the circumstances of every case being ordered in a number of groups a_1, a_2, a_3 . . ., and it is taken as *certain* that every character in one of these groups is necessitated by some character in each preceding group, and as *probable to a finite degree only* that it is necessitated by some other character in its own group (and therefore that it necessitates some other character in its own group) – that is what Lachelier's second principle amounts to – we tried to show that the accumulation of instances, however far it be pressed, can give only a finite probability that one character is necessitated by another. Such, then, is the situation with natural phenomena.

We find ourselves led to the conclusion that neither in the field of numbers nor in the field of nature, does eliminative probability, presented as the sole source of the validity of induction by repetition of instances, provide that type of induction with sufficient power for it to approximate to certainty. But in the repetition of instances lay the only hope that remained of raising to practical certainty the modest probability that is all that can be derived from deliberate elimination when applied to natural phenomena without any previous knowledge directing it and providing a guarantee. Thus we see that the theory which sees in invalidation the sole source, whether manifest or covert, of all induction has been found incapable of conferring an infinite probability on any law whatever. It does not allow physics to go beyond a modest probability, fixed *a priori*, however much care and pertinacity be devoted to the task.

We cannot reject this result absolutely, in view of the profound ignorance that there still is about the nature of induction and therefore also about its power and limits. All the same, we can accept it only if there exists no other plausible theory which permits us to avoid it. We have so far examined only one theory, one which, despite its advantages, could not fail to be hazardous. It was in fact willing to allow only one of the two elementary operations which reason can distinguish in the relation of facts to laws. It seeks to base even the confirmation of a law by its own instances on the invalidation of competing laws. That doctrine, which reason unconsciously embraces at first, leads only to a modest result and one not favourable to the natural sciences. It is time to free ourselves from the spell of its reputation by observing that if we had examined it before following it, we should have recognized it as something extreme, indeed more or less a gamble.

The other way must therefore also be tried. It is possible, and indeed natural, to conjecture that the confirmation of a law by its instances, induction by repetition, possesses a strength which comes not from the eliminative probability that is included in it, but from elsewhere. That is what we must do if we are to try, in Plato's words, to 'save' our knowledge of nature and open up at any rate an approach to certainty.

But such an idea takes us along a road that is still entirely unknown. It is in fact the renunciation of the doctrine that has governed, with a varying degree of distinctness, the thinking of logicians up to now. It is to turn away altogether from the direction in which our mind first goes in search of an explanation of the confirmatory power of instances, which culminated in Mr Keynes's analysis. We must look elsewhere for an explanation and analysis of it; for we cannot pretend that the power is self-explanatory. As we remarked at the beginning, the invalidation of a law by a counter-instance is intelligible in itself. The mind asks no questions about its foundations or its degree. It sees in it a simple and decisive operation. On the other hand, the confirmation of a law by a favourable instance does not possess the same degree of perspicuity. Its validity remains obscure. The principle according to which the probability produced by a number of instances that grows to infinity would approximate to certainty can surely not claim self-evidence. It would require some proof. It now seems that we cannot derive from the classic works the slightest help in our search for such a proof. For the final upshot of them is the theory of probability by elimination; and we have seen that by that theory the principle itself stands condemned.

But a recent work offers us precisely what we are looking for – a justification of induction by repetition and a proof that it tends towards certainty when the number of instances increases to infinity. It is a work that we know already, in fact Mr Keynes's *Treatise on Probability* in which, however, as we have seen he maintains the theory that we have shown to be fatal to the principle itself.

Let us set out Mr Keynes's excellent theorems; let us disentangle them from a traditional philosophy whose definitive refutation they contain; let us then examine the proofs that the author gives of them, and show why one of them seems to us to be incorrect.

CHAPTER IV

Induction by Confirmation

Mr Keynes's theory rests on the fundamental axiom concerning the probability of the conjunction of two propositions; the simplicity of this foundation is remarkable.

The probability of the conjunction of two propositions

What is the probability that two propositions p and q are both true? The answer is often given that it is the *product* of their two probabilities. That, however, is not in general, correct – only in the special case in which the two probabilities are independent of one another, that is, where the news that one of the propositions is true would not increase or diminish the probability of the other. In all other circumstances it is clear that the probability of p and q together is not after all equal to the product of their separate probabilities. In fact, if q is a certain consequence of p, the probability of pq is that of p alone; and if q is a probable consequence of p, the probability of pq is even greater than that product. Conversely, if not-q is a certain consequence of p, the probability of pq is zero; and if not-q is a probable consequence of p, the probability of pq is still less than the product.

The probability of pq is thus not a function of the initial probabilities of p and q, but a function of the initial probability of p and the probability of q given p; again, for reasons of symmetry – for even if p and q relate to events, that which occurs later may be one that is known, i.e. given, first – it is a function of the initial probability of q and the probability of p given q; and, again as a matter of symmetry, that must be the same function. That function is as before, the *product* of the two probabilities.

Let us designate by x/y the probability of x given y. Let h be the initial data. We shall then have

$$pq/h = p/h \times q/hp = q/h \times p/hq.$$

That is the principle; it is, at the very least, enormously plausible.

We notice that it is a universal one and independent of any assumption or hypothesis whatever.

The justification of induction by repetition

Let p be a general proposition or law, where the initial data h at the moment considered may indifferently either include or not include already a certain number of instances. Let q be the proposition that the law p will turn out to be verified in a new instance E.

If the law p is true, q is certainly true also. We then have

$$q/hp = 1.$$

The principle $$p/h \times q/hp = q/h \times p/hq$$

then gives us $$\frac{p/h}{p/hq} = \frac{q/h}{1}$$

That is to say, the probability of the law before a verification is to the probability of the law after that verification as the probability of that verification itself, before it occurred, is to certainty.

For the verification q to render the law more probable, we see that it is necessary and sufficient

(*a*) that p/h should not be zero, i.e. *the law possesses some probability, however weak, independently of the verification;*

(*b*) that q/h be smaller than unity, i.e. the verification of q does not follow with certainty from what is known already.

This theorem justifies induction by repetition.(1) It further establishes that it does not have any deterministic premiss, that its force does not come from a probability of elimination, and that it does not require that it be at least probable that the instances vary. It thus manages to reverse the philosophical doctrine that we criticised earlier, of which Mr Keynes himself still remains a supporter. Let us pause a moment to make these important consequences clear.

Induction by repetition does not have a deterministic premiss

We have just shown, in effect, that every verification which is not certain in advance renders the law more probable, on condition only that the law possesses already some probability, however slight,

(1) Once again, we must not exaggerate the scope of this purely theoretical proposition. For its asserts that the accumulation of cases which verify the law renders its verification more probable in *all the cases* (and therefore *in any given case*); but not yet *in all the cases*

of being true. Let X and A be the characters which it connects. If the discovery of A in a case of X, where its presence could not have been foreseen with certainty from previous knowledge, is to render X *necessitates* A more probable, the only assumption that it is necessary to make is therefore that X *necessitates* A is already probable to some degree p which is not zero, but may be as weak as one wishes.

This assumption no doubt entails that the presence of some character necessitating A in every case of XA is probable at least to the degree p; for it is probable to that degree that X itself is such a character. But that is *all* that it implies. It in no way implies that the presence of such a character is *certain*.

Thus we see that induction by repetition requires only, in order to increase the probability of the law X *necessitates* A, that it be probable to some degree that A is determined.

But to show that this form of induction does not have a deterministic premiss, it is further necessary to show that it can confer on the law X *necessitates* A a *higher* probability than the initial probability of A's being determined.

That is easy. Thus the initial probability d of the presence of some character which necessitates A in every case of XA must be equal to or higher than the initial probability p/h of the law X *necessitates* A. We are free to suppose that it is only equal to it. It is enough to assume that we were sure that *only* X *could necessitate* A. This particular supposition is no obstacle to the application of the theorem: the probability p/hq of the law X *necessitates* A after the new verification q is then *higher* than p/h, its probability before this verification, and higher in the ratio of certainty to q/h, the probability of the verification q in the previous state of information h. But we are sure that only X can imply A; the initial probability d of A's being determined is therefore precisely equal to the initial probability p/h of the law X *necessitates* A. Thus the verification q confers on this law a higher probability than the initial probability of A's being determined. That proves what we set out to show, that induction by repetition does not have a deterministic premiss.

still unknown to us or *in any given one of these cases*, which is what is needed to justify genuine inductions. In fact, the proof we have given assumes that the cases that have been recognized as favourable still belong to the sum total of cases. It is no longer valid if we consider only the sum total of cases still unknown.

The strength of induction by repetition does not arise from eliminative probability

This follows immediately from the preceding proposition. For all that a perfect elimination would establish is that only X *could* necessitate A. The law X *necessitates* A would in that case inherit all the initial probability of A's being determined, but that is all; that result is the most that elimination could achieve. We have now supposed that it has been achieved, and we have shown that the first new verification can go beyond it. That cannot now be because the new verification has a chance of eliminating some character competing with X for the title of sufficient condition of A, since we are supposing that that elimination has already been carried out. it follows that the mechansim of confirmation by instances is not reducible to probability by elimination.

In particular, let us take the case where the existence of some other character inseparably tied to A is probable to the degree p, and where we possess two instances of XA having no other common character. These two instances would exemplify what Mr Keynes calls a perfect analogy. Allowing himself to be guided by the doctrine of elimination, he judges that no new instance can in such a case add anything to the probability of the connexion X and A.[1] But his own theorem proves just the opposite; for it proves that the verification of the connexion in a third instance, provided only that it could not have been predicted with certainty, would render the law more probable than it was before. Admittedly, we cannot ever be sure that two instances of XA differ in all other respects, and indeed we are most often sure that if we searched thoroughly, we should find other resemblances. But it remains true that the probable elimination of these resemblances is not the only origin of the favourable effect of further instances; for even if we supposed that elimination achieved, new instances could continue to strengthen the law.

A new instance identical with one already obtained can render a law more probable

Let us imagine a universe in which two instances could be two in number without differing in any of their characters. The supposition

(1) *A Treatise of Probability*, page 226.

is unreal and even absurd. However, it may serve to illustrate a thesis. Mr Keynes himself uses it for this purpose when he asserts that 'if the new instance were identical with one of the former instances (and if this identity were known to us with certainty) a knowledge of the latter would enable us to predict it.' (*Ibid.*, p. 236.) It follows that we should learn nothing from the second, and so, by the theorem, the probability of the law would not be increased. We too may therefore be allowed to have recourse to the fiction of two identical instances in order to deny precisely what Mr Keynes affirms about them.

Let us begin by making clear what the mutual inferability of two identical instances would consist in. If we had established that the second case reproduced, along with X, all the characters in the first case other than A, we could not fail to be certain that it reproduced A also, even before it had been established that it did.

But it is obvious that such certainty is in no way different from *the actual certainty of A's being determined.* For what we should be certain of is that the total character consisting of the conjunction of all the characters which accompany A in one of its instances necessitates A, and cannot be reproduced without A's also being reproduced.

No doubt that is how things are in our universe, and it is not even an assumption; but only for the stronger reason that such a total character could not in fact be reproduced. On the other hand, in the fictional universe that Mr Keynes and we imagine ourselves in – he in order to affirm that two identical instances were inferable from one another, and we in order to throw doubt on it – the obstacle that arises from the absence of indiscernibles no longer exists, and the assumption becomes a real one.

Now it does not figure among the assumptions of the theorem under discussion. The assumptions there, we remember, are solely that X *necessitates* A possesses some initial probability, however slight. We are therefore free to suppose that we have a case where it is not certain that the character A is determined. In that case it would not be certain that an instance of X which exhibited all the characters of a previously known instance of XA other than A, exhibited A also. In virtue of the theorem, to establish the presence of A in the second instance would increase the probability of X *necessitates* A, as it would be a new verification that could not have been predicted with certainty.

That is not all. Let us suppose that it is actually certain that A is determined. The discovery of an instance of XA identical with an instance already obtained could still increase the probability of X *necessitates* A, and that in virtue of the very theorem of Mr Keynes which at first seems to have the opposite consequence.

Let us symbolize by L, M, N ... the characters in the two instances *other than X and A*. Since the first instance is one which is known, it is certain that XLMN . . . necessitates A; to establish that A occurs *with* XLMN . . . in the second case in fact tells us nothing. But if we establish that LMN . . . occur *with* X in this same case, we do learn something; it renders the law X *necessitates* LMN, of which it is an instance, more probable. As it happens, the verification of that law in the second case does not follow with certainty from its verification in the first. Otherwise it would be certain that *all* the cases of XA are identical, and a single one of these cases would suffice to render X *necessitates* A certain; and that supposition would make any new instance pointless.

Without that supposition, which is surely too unnatural to detain us, an instance of AXLMN . . . identical with a previous one will increase the probability of the law X *necessitates* LMN . . . when that law is not recognized to be impossible. We have taken it as certain that any X, if it is LMN . . ., necessitates A. The second instance of AXLMN . . . renders it more probable that every X is LMN . . ., so long as the contrary is not rendered certain by the discovery of an X which is not LMN . . . . A second instance of AX, identical with the first, would thus render it more probable that X *necessitates* A; and that would be so, *even if it were certain that A is determined.*

Mr Keynes's theorem thus has a consequence that is precisely the opposite of the one that he himself thinks that he derives from it, falling into error as he does through the traditional conception of the mechanism of induction. Far from implying that two instances known to be identical count only as one, it implies quite the reverse.

It is true that there are no identical instances, and indeed there cannot be any. Mr Keynes's assertion served only to illustrate the doctrine that, in regard to the number of instances, it is *only their variety*, certain or probable, that has any effect. In the same way, we have just illustrated by the same fiction the contrary thesis that, in regard to the number of instances it is *not only* their variety, certain or probable, that has an effect. And we have shown that this is actually a consequence of Mr Keynes's theorem.

State of the question

In the first part of this study, we convinced ourselves that the corroborative effect of collections of instances of a law cannot derive all its strength from eliminative probability if inductive knowledge of laws of nature, and, secondarily, laws of number, is to approach certainty. The previous theorem establishes that that condition is in fact satisfied. It ensures that induction by simple enumeration is not governed by the conditions of eliminative induction, and is capable in principle of improving on the maximum result attainable by the latter. The question of approaching certainty by accumulating instances, in the very conditions in which the doctrine of the possibility of elimination rendered such an approximation impossible, is thus reopened. But the question is not yet favourably resolved; it has not yet been proved that the multiplication of instances of a law will confer on it a probability that is capable of reaching and going beyond any fixed value.

Mr Keynes thinks that he can prove that also, but with the help of a special postulate.

Two necessary and sufficient conditions of the probability of a law tending towards certainty through the multiplication of instances of it to infinity

It is first of all necessary that the law should possess, at the outset, a probability higher than zero, however slight it may be ; we recognize in this the condition that we know already to be necessary if instances are to yield an increase in probability. But if that probability is to be carried beyond any limit by an infinite number of instances, *it is further necessary that, supposing the law to be false, its verification successively in an infinite number of instances should be infinitely improbable ; or, in more precise terms, that its improbability should exceed any limit for a sufficiently large number of instances.*

Let us suppose that the law is verified in all known instances, and that these instances are infinite in number. Either the law is true or it is false. The fact that it has been verified in all these instances would have been a certain consequence of its truth. If we allow, in accordance with the two conditions previously mentioned, that the falsity of the law would render this same fact infinitely improbable, and again that its falsity is not infinitely probable in itself, it follows that this fact, once established, renders its falsity infinitely improbable. And that condition, which is sufficient, is also necessary.

All this is a direct result of the axioms of probability. Let p/h and $\bar{p}/h$ be the respective probabilities of the truth and falsity of the law p in the state of information h from which we start. Let V be the fact that the law is verified in an infinite number of instances not included in the data h. V/hp and $V/h\bar{p}$ are then the respective probabilities of V, in the state of information h, on the two assumptions of the truth and falsity of the law p. V/hp is obviously equal to unity. We are looking for the probability $p/h\mathrm{V}$ conferred on the law p by the establishing of the fact V.

We have

$$(1)\ \mathrm{V}/h = p/h \times \mathrm{V}/hp + \bar{p}/h \times \mathrm{V}/h\bar{p}$$

In fact the probability of V in state of information h subdivides into the respective probability of V in each of the two exclusive and exhaustive alternatives p and $\bar{p}$, multiplied by the probabilities of these alternatives themselves; that is a fundamental proposition of the logic of probability.

Now the principle adopted at the beginning of the whole argument gives us

$$p/h \times \mathrm{V}/hp = p\mathrm{V}/h = \mathrm{V}/h \times p/h\mathrm{V}$$

(1) then becomes

$$\mathrm{V}/h = \mathrm{V}/h \times p/h\mathrm{V} + \bar{p}/h \times \mathrm{V}/h\bar{p}$$

from which we derive

$$p/h\mathrm{V} = 1 - \frac{\bar{p}/\ h \times \mathrm{V}/h\bar{p}}{p/h + \bar{p}/h \times \mathrm{V}h/\bar{p}}$$

$$= 1 - \frac{\mathrm{V}/h \quad p/h \times \mathrm{V}/h\bar{p}}{p/h + \bar{p}/h \times \mathrm{V}/h\bar{p}}$$

Thus we see that, if $p/h\mathrm{V}$ is to tend towards unity as the number of verifications included in V grows to infinity, it is necessary and sufficient that p/h should not be zero, and that V/hp should tend towards zero.

Replacement of the second condition by a condition that is only sufficient

Mr Keynes substitutes for the condition $V/h\bar{p} > O$ a stronger condition that implies it but is not implied by it, whose satisfaction he thinks he can guarantee.

The rule for the conjunction of two probabilities, applied from each case to the next, gives as the conjoint probability $V/h\bar{p}$ of n verifications of $x_1, x_2, x_3, \ldots x_n$ in the state of information h, and assuming the falsity of the law p, the value

$$V/h\bar{p} = x_1, x_2 \ldots x_n/h\bar{p} = x_1/h\bar{p} \times x_2/h\bar{p}x_1 \times \ldots \times x_n/h\bar{p}x_1x_2 \ldots x_{n-1}$$

that is, the probability of n successive verifications is equal to the product of all their probabilities, given in each case the prevous probabilities.

The factors of this product are all less than unity. For their product to tend towards zero in proportion as their number increases, it is plainly sufficient that they should not tend towards unity but remain lower than a fraction f, itself less then unity; that is to say that there should exist a finite quantity ε such that we have, whatever n may be,

$$x_n/h\bar{p}x_1x_2 \ldots x_n - 1 < 1 - \varepsilon$$

That is the condition that Mr Keynes goes on to try to satisfy.

We notice that it is sufficient but no longer necessary; for the product of a number of fractions that increases may tend towards zero even though its factors tend towards unity, as we can see from the product

$$\frac{1.2.3}{2\ 3\ 4} \cdots \frac{n}{n+1}$$

whose last term $\frac{n}{n+1}$ tends towards 1, even though the value of $\frac{1}{n+1}$ tends nonetheless towards zero.

We shall therefore next try to establish not only that, assuming the falsity of the law, its verification in an infinite number of instances is infinitely improbable, but also that, on that assumption, even if an infinite number of verifications were obtained, that would

not render it infinitely probable that the next instance would still verify the law.

As we have just seen, what we have now is not, as before, a necessary condition. It is doubtful, even at a first glance, whether our universe satisfies it. It asserts, in fact, that if we *know* that a rule has some exception, the observation of as many million or thousand million successive verifications as we like cannot result in the falling below a certain limit of the chance that the next instance will be actually an exception. It claims that if we have once seen a man more than three metres high, the observation of as many men as we wish of less than three metres could not render it as probable as we wish that men of more than three metres are as rare as we wish; or again, if we had shown that two properties of numbers are not always associated, the observation of their association in the case of all the numbers that have been tried, as large a multitude of them as we wish, could not render it as probable as we wish that they will be associated again in the case of the next number that we try. That is the condition that Mr Keynes requires if the accumulation of instances in the absence of any exception or contrary proof is to be able to render it as probable as we wish that every man is less than three metres tall, or two arithmetical properties are associated. It will be accepted that the condition must be a difficult one to satisfy. However, he thinks that he can fulfil it, and at the same time fulfil the fundamental condition that there exists an initial non-zero probability in favour of p, with the aid of a highly acceptable postulate that he calls the postulate of limited independent variety.

The postulate of limited independent variety

The postulate consists in the assertion that the characters in the universe under consideration are divided up into groups, a certain member of which necessitates the others, and that the number of these groups is *finite*. We find in Mr Keynes's work an extremely interesting discussion of the character and scope of this postulate.

It satisfies the first condition

It is a consequence of it that a character X taken at random possesses a finite chance *a priori* of necessitating the character A, also chosen at random. Thus, the character A possesses a finite chance of belonging to one or more groups chosen at random, since

the number of these groups is finite. It, therefore, possesses a finite chance of belonging to a group or groups which X belongs to, that is, of being present in all the instances of X. Thus the first condition will in fact be fulfilled[1].

But does it also satisfy the second? Mr Keynes's reasoning

It is the second condition that causes difficulty. Let us analyse Mr Keynes's argument as he sets it out, in a fairly condensed form, in his *Treatise* (p. 254).

The number of individuals or instances in the domain concerned – numbers or natural phenomena – can be thought of as infinite, and indeed must be infinite, since we are considering what the probability of a law becomes at the limit when the series of instances is indefinitely prolonged. *The number of characters* of these instances, and indeed of any one of them, may also be infinite. All that is finite is the number of *groups of characters* necessitated by a certain member of the group, or in other words *the number of characters which are sufficient to determine all the others.*

In consequence, the number of *distinguishable*, *non-identical instances is finite*. It is in fact limited by the number of combinations of the determining characters.

That is what Mr Keynes takes as his foundation.

If the law p is false, he says, it is false in at least one instance. But the number of distinguishable instances is finite: let their number be N. On the other hand, we naturally accept in virtue of the principle of equal ignorance, or indifference, that at any moment, it is no more probable that the instance that is about to turn up is one rather than any other of the instances that there are. It follows that the instance or instances that falsify the law p have, at any particular moment, a chance of turning up at least equal to $\varepsilon = \frac{1}{N}$, and we have, whatever n may be,

$$x_n/h\bar{p}x_1x_2 \ldots x_{n-1\,n-1}\ 1 - \varepsilon$$

(1) Strictly speaking, it would be necessary for this purpose to assert not only that the number of groups of associated characters is some finite number x, but also that there is a finite probability that this number x is less than a given number – for example, ten million. For if all finite numbers have the same chance of being that number x, it is infinitely more probable that x is higher than that it is lower than any assigned number, with the result that the consequence that we have just drawn from x's finiteness would no longer follow.

This reasoning rests on an unacceptable assumption

The nerve of this argument is clearly the finiteness of the number of instances. From this finiteness it must follow that there exists a finite lower limit to the probability that the next instance, supposing it to be taken at random, is one of the exceptions to the law. Now what has been given as finite is not *the number of individual instances*, but solely *the number of non-identical or distinguishable instances*, which we might call *the number of species*, meaning by that *infimae species*. Thus Mr Keynes's argument takes it for granted that instances that are not distinguishable from one another, however many there are of them, have no more effect than a single one and need not be separately considered.

Now that assumption is so strange that we should hesitate to attribute it to him, were we able to doubt that his argument requires it.

In fact, it amounts to saying that the proportion of individuals met with of different species cannot give any indication of the rareness or frequency of these species among the individuals of a genus. If, in a certain genus, we have met with individuals only of certain species in whatever number it may be, with only one exception, that fact cannot render it more probable than it was *a priori* that these species occur frequently in the genus, and therefore that a new individual of the genus, taken at random, will also prove to belong to those species. Experience would not be able to modify our initial ignorance of the relative importance of the species that exist, or the chances that there are that an unknown instance of a genus belongs to one rather than another of its species. Observation would not be able to teach us anything *de multis et paucis*. This assumption is really unacceptable and yet it is indispensable to Mr Keynes's argument.

If the assumption is not in fact made, the argument collapses. For we have assumed that the law p, 'X necessitates A', is false. There is, therefore, at least one combination of characters, at least one species, in which X is to be found without A; and the number of species is finite. It is, therefore, true that the probability, when we select a *species* of the genus X, under the conditions in which they are all equally probable, of chancing on a species that lacks A is at any moment greater than a finite value ε. *But with each observation what we directly select is not a species but an individual*; and for the argument to continue to have application, it would be necessary that it should be equally probable at any moment, *not that we should select any particular one of the individuals of the genus* X, *but rather an individual of any one of the species of the genus* X, *as any other*. But commonsense seems to say that if, when we select individuals of the genus X, we always come across members of species containing A, and that happens through as long a series as we wish, that makes it extremely probable, if not as probable as we wish, that within the genus X the species without A are rarer than the species containing A, if not as rare as we wish; and that, therefore, the individual which is about to turn up will also be of a species containing A, *precisely because* our ignorance leaves all the still unknown individuals of the genus A the same chance of turning up.

The argument that Mr Keynes built upon the postulated finiteness of the number of species rests, therefore, on the assumption, which is plainly contrary to the facts, that experience in no way alters the initial ignorance which makes us regard an unknown individual of a genus as having no greater chance of belonging to certain species of that genus than to others. We may see in the fact that Mr Keynes allowed himself to be seduced by so ill-based a construction the effect of his general doctrine of the necessary diversity of useful instances. Although we have already seen it to be false, it is a tenable view so long as it is the probability of laws that is in question, i.e. the existence or non-existence of certain species, but it falls into absurdity when we pass on to the rareness or frequency of the species that exist. We saw when we were studying his first theorem, that Mr Keynes remained attached to a philosophy of induction that is incompatible with his positive theory. But when we come to his second theorem, that philosophy enters into the proof of it in an extremely unfortunate way, and vitiates it.

Present state of the problem

We think that we have shown that if elimination is the only sphere of competence of induction, as writers on the subject and commonsense itself tend to believe, no induction in favour of a law can go beyond a moderate probability. We think that we have also shown that elimination is not the only field of operation of such inductions, and that the instances of a law have a corroborative power that does not derive from it and does not require a deterministic assumption. Finally, we think that we have shown that it has not yet been possible to prove that those instances when multiplied to infinity, can raise the probability of the law above any limit. That seems to us to be the present state of the logical problem of induction.

Annex

by R. F. HARROD

On the negative side, Nicod, rightly, rejects *a priori* assumptions that might be deemed to help to establish the validity of induction, namely the uniformity of nature, the principle that every event must have a cause, an *a priori* preference for simple laws, and the egregious 'principle of indifference' (viz the principle that, if there are n possibilities in the case, there is an equal probability $\left(\frac{1}{n}\right)$ of any one being true). He considers Keynes's proposal that induction could be justified by the assumption of a finite number of ultimate generator properties which would render possible the hypothesis of initial prior probabilities of various propositions, as required for a Bayes-type induction. Nicod affirms that to help to validify induction, it would be necessary to assume, not merely a finite number, but a definite number, of ultimate generator properties. Such an assumption would clearly be entirely unwarrantable. This point was put to Keynes by F. P. Ramsey, whether after correspondence with Nicod or not I do not know, and Keynes admitted that it had force.

On the positive side Nicod was convinced that the *ultimate* basis of induction was what could be inferred from the simple repetition of instances. More sophisticated kinds of induction, such as induction by elimination, require some prior knowledge (of probabilities), but induction, not depending on any prior knowledge, can rest only on simple enumeration. While convinced that this was true, Nicod did not think that he had found how to justify induction by simple enumeration.

After first reading Nicod I continued to ponder over this problem. And then the answer seemed suddenly to break in upon my mind. A first preliminary sketch of this answer was given in an article on 'Memory' (*Mind*, January, 1942). It was much more thoroughly developed in my *Foundations of Inductive Logic* (1956, Macmillan).

Perhaps I may be allowed to set out here in some very brief paragraphs the quintessentials of my solution.

It struck me that the validity of induction depends on an *anterior* principle, which we may call the 'principle of experience'. Let us first brush from our minds the problems relating to confirming or invalidating instances.

To assert that a proposition X has probability (a fraction between 0 and 1) means that for every time that one asserts a proposition the evidence of which is of a logical character *identical* with that of the evidence for X, one will be right np times for every $n(1-p)$ times that one is wrong.

If a traveller over a continuity of uniform character and of entirely unknown extent continually asserts that the termination of this continuity will not occur in a certain time, defined as a qth part of the time for which he has already travelled, where q is a very small fraction, he will be right much more often than he is wrong. This property of his assertions, taken as a whole, which can be predicted of them with *absolute certainty*, conforms to the definition of what is meant by saying that X is 'probable'. Nonetheless so far we have not got a satisfactory vindication of the principle of experience, since at any point of time some of the 'right' answers of the traveller will have already been given, whereas what he is interested in (as we are in making inductions) is the fraction of *future* answers that will be right. This point is dealt with in my book, I hope satisfactorily.

It is to be noted that the principle of experience, thus defined, is not an assumption, but an *a priori* certainty.

Now let us return to the instances of the concomitance of the property X with the property A. In the case of any particular X and A the instances that we get may not be representative of what is outside our experience. But if we take a large number of concomitances, viz of X_1 with A_1, of X_2 with A_2 ... of X_r with A_r ... where X_1, X_2 etc. are different properties, and also A_1, A_2 etc., we can assert of these concomitances taken together one or other of three things, one of which must be true.

1. The sample of samples (of concomitances of the various properties) lying within our experience has been a fair one: viz. the samples showing more uniformity than exists among the unknown phenomena is balanced, according to some 'normal error law', by those showing less uniformity.

2. The sample of samples has been biased towards showing